Regulating food law

Also available in the 'European Institute for Food Law series':

European Food Law Handbook
Bernd van der Meulen and Menno van der Velde
ISBN 978-90-8686-082-1
www.WageningenAcademic.com/foodlaw

Fed up with the right to food?
The Netherlands' policies and practices regarding the human right to adequate food
edited by: Otto Hospes and Bernd van der Meulen
ISBN 978-90-8686-107-1
www.WageningenAcademic.com/righttofood

Reconciling food law to competitiveness
Report on the regulatory environment of the European food and dairy sector
Bernd van der Meulen
ISBN 978-90-8686-098-2
www.WageningenAcademic.com/reconciling

Governing food security
Law, politics and the right to food
edited by: Otto Hospes, Irene Hadiprayitno
ISBN: 978-90-8686-157-6; e-book ISBN: 978-90-8686-713-4
www.WageningenAcademic.com/EIFL-05

Private food law
Governing food chains through contract law, self-regulation, private standards, audits
and certification schemes
edited by: Bernd van der Meulen
ISBN: 978-90-8686-176-7; e-book ISBN: 978-90-8686-730-1
www.WageningenAcademic.com/EIFL-06

european food law association
association européenne de droit alimentaire

Regulating food law

Risk analysis and the precautionary principle as general principles of EU food law

Anna Szajkowska

Wageningen Academic
P u b l i s h e r s

ISBN: 978-90-8686-194-1
e-ISBN: 978-90-8686-750-9
DOI: 10.3920/978-90-8686-750-9

ISSN 1871-3483

Cover: www.bartoszstefaniak.com

First published, 2012

© Wageningen Academic Publishers
The Netherlands, 2012

Table of contents

Abbreviations

ADI	Acceptable daily intake
AG	Advocate General
BR	Better regulation
BSE	Bovine spongiform encephalopathy
CAC	Codex Alimentarius Commission
EC Treaty	Treaty establishing the European Community
EC	European Commission
ECJ	European Court of Justice
ECR	European Court Reports
EEA	European Economic Area
EESC	European Economic and Social Committee
EFSA	European Food Safety Authority
EFTA	European Free Trade Association
EU	European Union
FAO	Food and Agriculture Organization
FSA	Food Standards Agency
GATT	General Agreement on Tariffs and Trade
GFL	General Food Law
GM	Genetically modified
GMO	Genetically modified organism
JRC	Joint Research Centre
NFR	Novel Food Regulation
NRC	National Research Council
OECD	Organisation for Economic Co-operation and Development
OJ	Official Journal of the European Union
OLFs	Other Legitimate Factors
RASFF	Rapid Alert System for Food and Feed
REACH	Regulation on Registration, Evaluation, Authorisation and Restriction of Chemicals
SCAN	Scientific Committee on Animal Health
SCF	Scientific Committee on Food
SCFCAH	Standing Committee on the Food Chain and Animal Health
SCTEE	Scientific Committee on Toxicity, Ecotoxicity and the Environment
SEA	Single European Act
SPS	Sanitary and phytosanitary
SSC	Scientific Steering Committee
TBT	Technical barriers to trade
TFEU	Treaty on the Functioning of the European Union
UNRISD	United Nations Research Institute for Social Development
USDA	United States Department of Agriculture
vCJD	Variant of Creutzfeldt-Jakob disease
WHO	World Health Organization
WTO	World Trade Organization

1. General introduction

1.1 Introduction

A quiet legal revolution has taken place in the EU. Its key moments were so subtle and behind the scenes that one might not even have noticed. Looking around, even an eagle-eyed viewer will note that not much has altered in the elements of legal landscape. And he will be right: it is not the landscape, but the ground we stand on that has changed. When I dot the last i's and cross the last t's of this book, food safety law in the European Union has – almost to the day – been functioning for one decade within the framework of the fundamental principles of risk analysis and precaution.[1] This study focuses on a legal analysis of these principles and their meaning for EU food safety law. By way of introduction, however, the historical, political and sociological context of the momentous restructuring of EU food law is sketched.

'Food shall not be placed on the market if it is unsafe' – states Article 14(1) of Regulation 178/2002 (the co-called General Food Law – GFL).[2] This Regulation set up the foundations for reformed food policy by laying down the general requirements of food law, establishing the European Food Safety Authority, and introducing procedures for handling food safety matters. All food-related undertakings, whether for profit or not; public or private; at any stage of production, processing and distribution; holding for the purpose of sale or other form of transfer, whether free of charge or not, must comply with the stated requirements.

Food is considered unsafe if it is:
- injurious to health; or
- unfit for human consumption.[3]

The concept of 'injurious to health' is linked to safety; the term 'unfit' is associated with unacceptability. Food can be rendered unfit, e.g. by the presence of a foreign object, such as a dead fly in a soup, or unacceptable taste or odour.

If a local farmer has a small vegetable plot and wants to sell carrots, in which the level of lead is 0.11 mg/kg instead of the 0.10 mg/kg permitted by EU law because the plot happens to be too close to a busy road, his food will be found in breach

---

[1] The law was published on 1 February 2002 and entered into force on the 20th day following that of its publication.
[2] OJ 2002, L 31/1.
[3] Art. 14(2) GFL.

of the legislation setting maximum levels for certain contaminants.[4] The carrots will be presumed to be unsafe for the purposes of Article 14 GFL.[5]

Lead present in carrots (or other foodstuffs) is a contaminant – a substance not intentionally added to food, but present as a result of cultivation practices, production processes or environmental contamination.[6] EU legislation stipulates that food containing a level of contaminant unacceptable from a public health viewpoint cannot be placed on the market. Because many contaminants, such as aflatoxins, heavy metals and dioxins, occur naturally, the legislator cannot ban them totally. Instead, it is possible to establish for these substances levels that are as low as can reasonably be achieved.

The aim of Article 14 GFL is to protect public health. The scope of application of the requirement is very broad. The farmer cannot make a carrot cake and bring it to a buffet at a dance. Nor can he offer these carrots for free on a local market, even if he puts clear information that they exceed the permitted level of lead by 0.01 mg/kg and that eating them is at the consumer's own risk. Moreover, suppose somebody is particularly interested in buying carrots containing lead at a level higher than set out in food safety standards because he or she believes that the higher the level of lead, the better the food. If that person contacts the farmer and offers to buy the whole yield of the carrot plot for personal use, the farmer would still not be allowed to sell the carrots.

While it may seem strange that a consumer would specifically want to buy food contaminated with lead, food safety requirements may collide in this way with alternative medicine. Products that cannot be sold as medicine because they have not been approved as such cannot be sold as food either if their active (or other) substance renders them unsafe under Article 14 GFL.

The regulatory regime established by the General Food Law is strict. Food found unsafe falls under the small and exceptional category of *res extra commercium* (a thing outside commerce). Some other examples of *res extra commercium* are weapons, human body parts, illegal drugs, and *res publicae* (public waters, etc.). The overriding public morality, safety and welfare interests, such as the protection of human life and health, makes these objects condemned by society and unsusceptible to being traded.

---

[4] Reg. 1881/2006, OJ 2006, L 364/5.

[5] In some cases, an assessment of the harm that might be caused and its likelihood should still be carried out, taking the factors in Article 14(3) and (4) GFL and the specific legislation in breach into account. E.g. maximum residue levels (MRLs) established in legislation are indicative of good agricultural practice. Exceeding MRLs, although considered an offence, does not immediately mean that a food is unsafe in the sense of Art. 14 GFL (EC, 2010; FSA, 2007).

[6] Art. 1(1) Reg. 315/93 laying down Community procedures for contaminants in food, OJ 1993, L 37/1.

Policy makers have to decide on the level of risk that is acceptable to consumers. In practice, it means that decision makers limit consumers' freedom of choice when it comes to their safety (Grigorakis, 2006; Rippe, 2000). Referring to the carrot example, the question remains whether the consumer should be free to choose products that are considered unsafe. The unsafe carrots can be banned, but they can also be accompanied by all available information concerning possible risks to human health so that the consumers can decide themselves. Are they able, however, to balance all risks and benefits and to make informed choices? Which risks should be handled by the consumers and which should be entrusted to experts? Which experts should decide what risks are acceptable for society? What to do when different experts have contradictory opinions on the same subject? These questions are core issues of risk regulation and this study.

1.2 Risk society

Ulrich Beck, one of the most cited sociologists, proposes the concept of risk society to describe the manner in which modern society living on a high technological frontier responds to risks (Beck, 1992). The consequences of scientific and technological development are risks. These risks, such as environmental pollution or newly discovered technologies, are 'manufactured risks' – produced by human activity (Giddens, 1999).

Risk society is characterised by societal changes that are referred to as reflexive modernity (or post modernity). The argument is that – while industrial society was shaped by the production and distribution of wealth – risk society is dominated by the distribution of risks resulting from wealth production (Figure 1.1). The concept of risk society changes the way social hierarchies and classes are constructed. Modern risks may easily go beyond boundaries of time and space. They can affect future generations and are no longer confined to national borders. Hence, although wealthy classes may still be more able to balance certain risks, for example by moving to more expensive residential areas located further away from industrial zones, according to Beck, the ability to avert risks is actually more related to an individual's ability to access knowledge or information on risks, rather than to wealth.

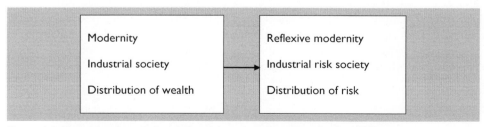

Figure 1.1. Transition from industrial to risk society (based on Beck, 1992).

1.2.1 Risk versus hazard

The distinction between risk and hazard is important in the risk society context. A *hazard* is something (object, substance, activity, phenomenon) that has the potential to cause adverse effects. Breathing in asbestos dust is hazardous because it can cause cancer. For food, hazard is defined as 'a biological, chemical or physical agent in, or condition of, food or feed with the potential to cause an adverse health effect'.[7]

A *risk* is the likelihood (the chance or the probability) that a hazard will actually cause its adverse effects, together with the severity (the magnitude or a measure) of that effect (Figure 1.2). For example: 'The annual risk of a fatal (effect) food poisoning (hazard) is less than one in 100,000 (risk)'; or 'x per cent (probability) of smokers (hazardous activity) will develop lung cancer (effect)'. The likelihood can be described as probabilities or frequencies (e.g. one hundred cases per year) or in a qualitative manner ('significant', 'negligible').

Figure 1.2. Definition of risk.

1.2.2 Risk versus danger

The difference between risk and danger is less obvious, especially since both concepts are often used synonymously. Beck defines risk as a systematic and intentional way people deal with insecurities, whereas danger is a situation in which something unpleasant or damaging might potentially happen, regardless of our choices. In other words, risks are taken[8], whereas dangers happen to people (Beck, 1992).

Thus, the distinction between risk and danger lies in the role of the decision making process. In the case of dangers, a loss may occur irrespective of whether a decision has been taken or not (Luhmann, 1993). Luhmann also provides an example of a meteorite hitting Earth to show that modern society sees danger in the context of risk and considers danger seriously only as a risk. The probability of a meteorite striking the ground with disastrous effects is underestimated because we can do nothing to prevent it (Luhmann, 1993).

---

[7] Art. 3(14) GFL.
[8] Or rather, are avoided, as some authors note a shift from a risk taking culture to a precautionary culture (Pieterman, 2001).

Because manufactured risks are the result of human activity, societies can assess the levels of these risks and decide whether or not to accept them. Risks express, by their very nature, an anticipatory component: they signify a future that is to be prevented. According to Bernstein, the notion of risk, or even the foundation of probability theory by Blaise Pascal and Pierre de Fermat, introduced a major change in the perception of the future by society. The concept of risk allowed people to cross the boundary of time and decide on what may happen in the future, which was previously considered a whim of the gods (Bernstein, 1998).

Therefore, risk society is not necessarily more dangerous or hazardous than social orders existing before (Giddens, 1999). The distinguishing feature of reflexive modernity is that society is becoming increasingly preoccupied with possible (future) effects of its activity. The progress of science and technology turned many dangers into risks, creating a culture that lives 'in the future rather than in the past' (Giddens and Pierson, 1998).

Moreover, Beck points out that science and research do not contribute to safety. On the contrary, progress in science creates more questions and more uncertainties. This explains the paradox that food safety has become an issue just as food has become safer (Busch, 2004). Because defining what may happen in the future and choosing among alternatives lies at the heart of decision-making processes, it is inevitably risk-connected. Every decision – or non-decision – has a risk, which may only be higher or lower (Trute, 2003).

1.3 Food as risk

The area of food production and consumption does not escape the influence of newly 'manufactured' risks resulting from techno-scientific innovations. The BSE (bovine spongiform encephalopathy) outbreak, genetically modified organisms and animal cloning have brought food to the forefront of political debate. Although manufactured risks attract greater attention[9], conventional risks or dangers, such as food poisoning, microorganisms, bacteria or pesticides, are subject to similar mechanisms of reflexive modernity (Mol and Bulkeley, 2002). The changing character of the agro-food chain and the role of science in the policy-making process make society equally interested in these risks and create expectations that scientific experts will provide information necessary to limit the impact of both types of risk.

This confirms the trend of a shift from danger to risk. An increasing number of issues are seen as related to human behaviour, and thus no longer 'natural'. Trute notes that, in the case of a flood, it takes a few days to shift the perspective from a

---

[9] E.g. the analysis of the US Food and Drug Administration legislation shows that regulation for new synthetic chemicals is more frequent than for natural ones (Löfstedt, 2005).

natural catastrophe to a discussion about who is responsible for flood prevention and hazard management system (Trute, 2003).

Similar changes can be observed in the field of food safety. The traceability system established in the EU for food and feed at all stages of production in most cases enables the identification of the source of risk, hence to trace it back to human decisions. Because, in the concept of danger, a loss cannot be prevented by human decisions, dangers are outside the scope of regulation. The General Food Law does not distinguish between risks and dangers and refers only to physical, chemical or biological agents that can cause an adverse effect (hazards) and risks understood as a function of probability and severity of that effect.[10]

Busch indicates several transitions that have changed our understanding of food safety: transition from local to global markets, changes in what is shipped over long distances, changes in scale of production and processing, and new technologies (Busch, 2004).

1.3.1 Globalisation and changes in scale of production and processing

A quick look at the shelves of any supermarket suffices to confirm the obvious: food trade is truly globalised. Our diet does not depend on local products and does not have a seasonal character. This is not only true for people living in urban areas in industrialised countries; those in rural areas and low-income nations experience similar changes in their consumption patterns.

New technologies and social organisation have also led to an increase in the scale of food production. Large producers, processors and retailers dominate the food market. Nowadays, a farm woman selling a few birds on a local market will rarely be the first association with 'chicken trade' – a huge barn with thousands of birds is more likely to come to our minds. A modern high throughput poultry abattoir may slaughter 9,000 broilers per hour and 20 million per year. Similarly, fresh ingredients for a vegetable salad will rarely come from a vegetable plot nearby. Moreover, possibly they have never even touched soil. A big part of the vegetable supply available the whole year round in European supermarkets comes from Almería in southern Spain. The 'sea of plastic' or 'costa del polythene', as it is often called, is an area of around 40,000 hectares of greenhouses, the largest concentration in the world, located around 30 km southwest of the city. Over 2.7 million tonnes of produce are grown under polythene continuously from October to July. Most plants grow from bags filled with grains of white stone, into which chemical fertilisers drip from big, computer-controlled vats.

---

[10] However, other linguistic versions of the Regulation use various terminology, e.g. Dutch *'gevaar'*, German *'Gefahr'*, French *'danger'*, Italian *'pericolo'*, Polish *'zagrożenie'*, or Spanish *'factor de peligro'*.

The modern food chain has been lengthened greatly. Globalisation, as well as changes in production and processing, have created more abstract relations between the producer and consumer. Busch argues, '[t]hey have made us into consumers, whose role is confined to retrieve the goods, place them in carts and bring them to the checkout counters, with as little disruption during the shopping process as possible' (Busch, 2004). Consumers are the last, anonymous link, without any personal ties to producers, processors and distributors, and with little or no knowledge of what happened to the cornflakes between a corn field and the supermarket shelf.

Because it is not possible to infer production methods or ingredients from the appearance of food products, information asymmetries between producers and consumers are huge. Moreover, even if the extent to which food labelling does contain detailed information, consumers often are not able to evaluate it adequately.

From a competitive market perspective, the gap between what we know about food products and what we could know can lead to information failures, where the market itself does not supply enough information for buyers to evaluate competing products and make informed choices (Majone, 1996). For this reason, a number of government requirements concerning consumer protection legislation, food safety and labelling requirements aim to reduce this information gap and mitigate its consequences.

From a risk regulation perspective, however, labelling is also linked to a fundamental question of a choice between paternalism and autonomy or sovereignty of consumers. Two extremes related to consumers' freedom of choice consist of either banning a risky product or activity, or – instead of banning – giving consumers all available information concerning risk and leaving them the choice (Grigorakis, 2006).

1.3.2 New technologies

Industrial processing technologies are widely applied in the food chain to retard spoilage; improve taste or colour; and change composition, structure or nutritional value of foods. As a result of these technologies, we have little or no idea what ingredients are in food products. Hazards are coded in chemical formulas, contained in modified primary molecular structure or nanoparticles. The Eurobarometer survey of 2010 on food-related risks assessing the concerns of consumers across Europe shows that EU citizens feel much more confident about being able to personally take steps to avoid diet and health-related issues ($>70\%$), such as high fat or heart disease, than to deal personally with chemical contamination ($<40\%$) or new technologies ($<30\%$). Pesticides, antibiotics, pollutants like mercury and dioxins, and finally animal cloning are perceived by consumers as risks to be 'very worried' about (Eurobarometer, 2010).

Chemical risks and risks related to new technologies attract the highest level of concern because they escape our perception. Risk perception is the predominant method by which people evaluate risks. It is an instinctive, intuitive, fast, and mostly automatic reaction to dangers. Slovic *et al.* (2004) describe this *risk as feelings* as an 'experiential system that enabled human beings to survive during their long period of evolution'. Risk perception uses images and quick associations and is linked by experience to emotions and a feeling whether something is good or bad. It tells us whether to walk down a dark street, approach an animal or drink strange-smelling water.

Chemical and nuclear technologies, however, are not perceptible to the senses. Risks remain hidden behind chemical formulas or closed in the invisible ultra-small world of nanotechnology. Their harmful consequences are often delayed and rare, hence we have little experience and historical reference on how to deal with these new risks. The mechanisms underlying these complex technologies and activities, remaining outside the reach of consumers' knowledge, have stimulated the creation of *risk analysis* – an analytical system based on algorithms and normative rules, using probability calculus, formal logic and risk assessment (Slovic, 1987).

Because risk perception (*risk as feelings*) is experiential and intuitive and takes a broader array of factors into account (such as voluntariness, fairness, or controllability), laypeople's reactions to risks may differ from scientific evaluation of risks made by experts (*risk analysis*). In this context, Slovic *et al.* (2004) add a third element to the distinction between risk as feelings and risk analysis: 'When our ancient instincts and our modern scientific analyses clash, we become painfully aware of a third reality – *risk as politics*'. Probably in no other area of risk regulation is this third reality more evident than in food safety governance.

1.4 Food safety politics

Most of what has been said about reflexive modernity and risks can apply to a whole range of products and activities: pharmaceuticals, cosmetics, clothing, toys, automobiles, and electric home appliances just to name a few. However, food is a particular area of public policy for a number of reasons.

Obviously, because food is among *basic needs* of mankind, effective food safety is a policy area everyone has a stake in. The physiological need to eat every day makes access to healthy food crucial for every citizen's life and well-being. Food safety regulation affects us directly and continuously.

Eating is an important and integral part of our lives and – although varying depending on lifestyle and economic or social status – few people eat simply for nutrition. Food has also an important *cultural dimension*, and national and ethnical cultural heritage has a strong influence on the ways we use food.

Developments in microbiology, chemistry and food technology, but also rapid transport and improved handling and packaging techniques have decoupled production locations from consumption and added a global dimension to trade in foods. Divergent food safety requirements and food standards existing in national laws give rise to trade barriers. As food is among the most traded commodities in the world, these differences have a huge *economic impact* and trigger regulatory efforts at international level to limit governments to use food safety as non-tariff trade barriers.

This predominant model of a large scale, industrialised 'global food system' is often opposed to local food systems, related to local markets. Local markets are based on face-to-face ties between producers and consumers and are characterised by trust and personal relations (Hinrichs, 2000). As mentioned earlier, the lengthening of the food chain has increased uncertainties and resulted in the *risk* becoming an issue. This risk, however, cannot be evaluated by consumers individually. Not only because of market information asymmetry, but also because – even if all necessary information were available to make an informed choice – many consumers would still feel not competent to evaluate risks without an advanced knowledge of chemistry, biology, or biotechnology. The consumer transaction costs would simply be too high (Rippe, 2000).

Because consumers, being not able to evaluate risks themselves, are dependent on external knowledge, the regulation of food safety represents a crucial – and also very emotional – dimension of public policy. Few other areas are so *politically sensitive* as those concerning food safety (Ansell and Vogel, 2006b). The BSE crisis is the best example of a food safety regulatory failure that shattered citizens' trust in political institutions and led to one of the biggest political crises in the history of the EU (Ratzan, 1998).

Therefore, even if there are good reasons to believe that it is in the interest of producers and suppliers to sell only safe food, the task of ensuring food safety cannot be left to the market forces alone. The intensity of regulatory intervention to correct market failures and to protect interests such as health (referred to as 'social regulation' or 'risk regulation')[11] has rarely been questioned (Hood *et al.*, 2001; Joerges, 1997; Majone, 1996). Regulatory activities of modern states are

---

[11] To use Majone's distinction, there are three main functions of the government in the socio-economic sphere:
- income-redistribution function – transfer of resources from one social group to another, provision of some goods;
- stabilization function – preservation of satisfactory levels of macroeconomic growth, employment and price stability; and
- economic and social regulation – correcting information failures and market failures, such as monopolies, or regulating market mechanisms to protect important values, such as human health (Majone, 1996).

increasingly characterised by a focus on risk and risk is becoming one of the dominant forms of regulation in the EU internal market.

1.5 Food safety regulation[12] – historical outline

1.5.1 Market-oriented regulation

Although people have always been concerned with food safety, food regulation remained, until recently, mainly local or national and principally referred to commercial transactions, focusing on weights, measures and adulteration (Snyder, 2006). With one exception referring to the environmental law, EU regulatory intervention in the areas of health and consumer protection between 1958 until the end of the 1970s remained very limited (Majone, 1996).

Between the 1970s and 1980s, a growing body of legislation and case law concerning foodstuffs was developed in the European Economic Community as the European Union was then called. This stage was related to the development of the EU internal market and food legislation was aimed at eliminating barriers to intra-EU trade. This market-oriented phase was mainly characterised by a *vertical* approach to harmonisation. Legislative efforts were focused on issuing directives harmonising standards concerning specific products (compositional and technical standards).

The choice of detailed harmonisation as a way to remove obstacles to trade consisting in different national quality standards made the achievement of the common market very difficult. The European Courts' judgements on, i.e. pasta, jam, chocolate, feta cheese, beer, energy drinks, cornflakes fortified with vitamins and minerals, Scotch whisky, and finally *Cassis de Dijon* – French sweet liqueur made from blackcurrants, show the bumpy road to the internal market.

The challenge was due not only to the abundance of national regulations in the area of food safety, but also to the 'extreme sensitivity of public opinion in this field' (EC, 1985b). As mentioned above, our distinctive attitudes towards food – what and how we eat – are related to our identity and largely determined by cultural, religious and national backgrounds. Food policy in the EU had to reconcile the European patchwork of diverse national eating habits and cultures within a single

---

[12] Regulation here is understood in a broad sense, encompassing laws, regulations and administrative provisions, similarly to the definition of 'food law' in Art. 3(1) GFL. It has to be noted, however, that the term is increasingly used in a narrower sense, referring to prescriptive rules as long as they are not formulated by the legislature (primary law) or the courts (judgments). Regulation in the narrower sense is thus confined to bureaucratic and administrative rule making and does not relate to legislative or judicial rule making (Levi-Faur, 2010). From an American perspective, Selznick defines regulation as 'sustained and focused control exercised by a public agency over activities that are socially valued' (Selznick, 1985). In applying this narrower sense in the European context, however, one should rather speak of bureaucratic-administrative authorities, since regulatory regimes are rarely based on public agencies.

market – a task very difficult to achieve when most harmonisation decisions required unanimity in the Council.

The year 1985 marked the advent of a more ambitious plan for the European internal market, as set out in the European Commission's White Paper 'Completing the Internal Market' (EC, 1985a). The White Paper included about 300 legislative measures to be taken by 1992 to incorporate components that were still hampering the functioning of the internal market, but – importantly – the document also envisaged the removal of all physical, technical and tax frontiers to create an area where people and goods can move freely. The internal market was included in the Treaty by the 1986 Single European Act (SEA), which came into force in 1987. The SEA gave teeth to the internal market objectives by introducing a more effective decision-making procedure – qualified majority voting – for most areas concerned.

For food law, the envisioned completion of the internal market meant a shift from systematic harmonisation of national standards to *horizontal* rules setting out essential requirements of a broad application (such as labelling rules applying to all foodstuffs). This new approach was generally confined to approximation of laws in the area of health and consumer protection, settling for mutual recognition of national standards in other areas.[13]

Despite the introduction of the horizontal approach, food legislation in the EU remained a patchy 'spill-over' effect of the internal market goals (Vos, 1999), lacking a coherent structure. Interestingly, already in 1984, in one of the first publications on European food law, Krämer (1984) noted in this regard:

> [T]here are no precise rules regarding the level of safety; this lack is due to the fact that there is no across-the-board Community Directive applicable to the safety of foodstuffs in general ... The only valid solution would seem to be to formulate Community safety rules. The individual measures described should be integrated into coherent, binding, and acceptable measures at Community level ... Such an instrument could lay down general safety requirements for normal use and a foreseeable misuse. It could endeavour to integrate the different national safety acts into one coherent framework, striking a balance between the diverging interests of open frontiers and safety for consumers/users. An EEC-wide internal market with different concepts of safety and with substantially different safety requirements will, in the medium or long term, hardly be a realistic perspective.

Krämer considered the integration of food safety rules into the general EU rules an important (and necessary) challenge for the food safety policy. It took, however, nineteen years before general requirements for food safety applicable to all

---

[13] For the concept of mutual recognition see Chapter 2.

legislation were developed at EU level. The need for radical reforms became evident during the BSE outbreak – the most important food safety regulatory failure. The shortcomings of the existing fragmented food safety legislation revealed by the crisis prompted the next stage in the development of European food policy.

1.5.2 Tipping point: BSE crisis

Although the first cases of BSE (bovine spongiform encephalopathy) were already detected in cattle in 1986, the British government denied for a decade any possibility that 'mad cow disease' might cross the species and affect humans, despite the growing public concern over the health effects of eating meat of BSE-diagnosed cattle.[14] The crisis broke on 20 March 1996 when the UK officially confirmed that a possible link between BSE and a variant of the mortal Creutzfeld-Jakob disease (vCJD) infecting the human brain had been established (BSE Enquiry, 2000). The European Commission this time recognised the seriousness of the problem and – one week later – issued a ban on all exports of British beef.[15] The embargo, however, did not end the controversy concerning food policy and public distrust in the EU public institutions.

The BSE crisis brought to light darker secrets of the industrialised food production process. European consumers woke up to the fact that cattle had been fed animal materials on a mass scale, and – more shockingly – these animal products came from slaughtered cattle, which turned natural herbivores not just into carnivores but into cannibals as well. This discovery shaped the public debate about risks in the food chain and the role of science in effectively responding to them.

The attitude of the European public and growing anxiety about possible consequences of this maladministration led to serious political repercussions. In September 1996, a temporary committee of enquiry was established by the European Parliament to investigate the handling of the crisis.[16] The enquiry report presented in 1997 accused the Commission of protecting the market and business interests over public health (European Parliament, 1997). It also revealed lack of transparency of the decision-making process at EU level and called into question the relation between decision makers and scientific committees providing risk

---

[14] For more details on the BSE crisis see, e.g. Vos (2000a), Millstone and Van Zwanenberg (2001) and Grossman (2006). The BSE crisis is considered important because of political and societal reactions it provoked. Opinions about the magnitude of the human epidemic still remain divided. Some believe that the disease could have reached its peak already. Others, however, predict that thousands may still be affected in the future because of the long incubation period of the disease. It has to be noted, however, that the number of people directly affected by vCJD is still lower than in other food safety incidents.

[15] Dec. 96/239/EC of 27 March 1996 on emergency measures to protect against bovine spongiform encephalopathy, OJ 1996, L 78/47.

[16] OJ 1996, C 261/132.

assessment, as well as a lack of coordination and openness of these scientific committees.[17]

The European Parliament did not confine itself to pointing out the shortcomings of the European food policy. It also formulated recommendations for improving the EU food legislation and the Commission committed itself to set a major reform of the food safety policy in the EU in motion.[18]

1.5.3 Europe after BSE: integrated food safety governance

The report was shortly followed by a series of Community actions. As early as April 1997, the Commission Communication on Consumer Health and Food Safety was issued, followed by the Green Paper on the General Principles of Food Law in the European Union (EC, 1997a,b), and – a little over two years later – the White Paper on Food Safety (EC, 2000a). The policy reforms outlined in these documents were driven by the need to guarantee a high and uniform level of food safety across the EU through a comprehensive EU policy.

This policy was enshrined in Regulation 178/2002 – the General Food Law – adopted on the 28th of January 2002. Prior to the reform, food safety at EU level followed from other policies, such as agriculture or the creation of an internal market. The Regulation provided a framework for the development of an independent food safety policy in the EU.

A new policy is usually integrated in two ways: *directly* – by creating a set of principles, and *indirectly* – by establishing an institutional framework (Ugland and Veggeland, 2006). These are the two core elements of the food policy established by the General Food Law, introducing the precautionary principle and risk analysis as general principles of food law and creating the European Food Safety Authority (EFSA) as part of the scientific institutional framework (cooperating and networking with its counterparts in Member States).[19]

The General Food Law established an integrated EU food policy in its own right, but this approach can also be characterised as an attempt to create an EU scientific *governance* of food safety. Governance here is understood in a normative sense, referring to a model organising the decision-making process or a framework for

[17] Because the Scientific Veterinary Committee advising the European Commission was chaired by a British scientist, the Committee was under a significant pressure from the British Ministry of Agriculture, Fisheries and Food (MAFF).
[18] See Santer (1997).
[19] In addition, the GFL introduces a number of regulatory actions to bring coherence to the European legislation by harmonizing essential provisions concerning food products. It establishes, i.e. a traceability system, attributes primary responsibility for ensuring the safety of the food chain to businesses, and sets up rapid and effective procedures to manage food safety emergencies.

managing society.[20] This normative approach is outlined in the Commission's White Paper on European Governance (EC, 2001a), emphasising openness, participation, accountability, effectiveness, and coherence of policies. This approach to governance directs attention towards 'the problems to be solved and the processes associated with solving them, rather than the relevant agents' (Caporaso, 1996). This holds true for the EU political system, where regulation becomes a question of governance itself. As noted by Levi-Faur, 'much of the academic and public discussion of regulation nowadays deals with the governance of regulation itself (or regulating regulation) rather than governance via regulation. The growth in the scope and number of regulations raises issues of effectiveness as well as issues of democratic control' (Levi-Faur, 2010). This study follows this approach.

1.6 European Food Safety Authority

The creation of the European Food Safety Authority – a new independent institution serving as a scientific point of reference on all aspects of food safety and communicating the results of these risk assessments to the consumers – was the response to the Parliament's enquiry report pointing out the lack of openness, independence and coordination of scientific advice at EU level. Based on the principles of scientific excellence, transparency and independence, EFSA was designed to help restore consumer confidence in EU food safety policy after the BSE crisis.

Science sets everything in motion, but a decision about what risks are acceptable and what should be done to reduce them is taken at a political level. This separation of independent risk assessment from political risk management means that EFSA is not vested with any regulatory powers. Complex risk management requires an expert involvement in the policy-making process ('expertocracy'). Democratic governments, however, cannot be successful without public acceptance. As Beck pointed out, what matters most in risk regulation is therefore making the basis for decision-making transparent and publicly accessible (Beck, 1992). Article 38 GFL lays down basic principles of EFSA's transparency (see Textbox 1.1).

The role of EFSA in the 'democratisation' of risk regulation by informing the public about food safety and its institutional independence is thus an important element of the reformed scientific food safety governance in the EU. The accessibility of information for interest groups is a prerequisite for accountability, as officials or institutions cannot be held accountable without transparency (Dressel *et al.*, 2006).

---

[20] Asselt and Renn (2011: 443) apply the term governance specifically in the context of risk-related decision making, defining 'risk governance' as 'the critical study of complex, interacting networks in which choices and decisions are made around risks' (descriptive approach) and as 'a set of normative principles which can inform all relevant actors of society how to deal responsibly with risks' (normative approach).

Textbox 1.1. Article 38 of the General Food Law.

Article 38

Transparency

1. The Authority shall ensure that it carries out its activities with a high level of transparency. It shall in particular make public without delay:
   a. agendas and minutes of the Scientific Committee and the Scientific Panels;
   b. the opinions of the Scientific Committee and the Scientific Panels immediately after adoption, minority opinions always being included;
   c. without prejudice to Articles 39 and 41, the information on which its opinions are based;
   d. the annual declarations of interest made by members of the Management Board, the Executive Director, members of the Advisory Forum and members of the Scientific Committee and Scientific Panels, as well as the declarations of interest made in relation to items on the agendas of meetings;
   e. the results of its scientific studies;
   f. the annual report of its activities;
   g. requests from the European Parliament, the Commission or a Member State for scientific opinions which have been refused or modified and the justifications for the refusal or modification.
2. The Management Board shall hold its meetings in public unless, acting on a proposal from the Executive Director, it decides otherwise for specific administrative points of its agenda, and may authorize consumer representatives or other interested parties to observe the proceedings of some of the Authority's activities.

1.7 Science-based food law

The general principles and requirements of food law constitute an umbrella covering all food legislation at both EU and national levels and applying to the entire food chain. Articles 5-8 of the GFL set out the following principles of food law:

- Food law shall pursue the general objectives of 'a high level of protection of human life and health and the protection of consumers' interests, including fair practices in food trade, taking account of, where appropriate, the protection of animal health and welfare, plant health and the environment'; while aiming to achieve the free movement of food and feed in the EU (Article 5).
- Food law shall be based on risk analysis (Article 6).
- In specific circumstances, the precautionary principle may be invoked (Article 7).
- Food law shall aim at the protection of consumers' interests and shall provide a basis to make informed food choices by aiming at the prevention of fraudulent and deceptive practices, the adulteration of food, and any other practices which may mislead the consumer (Article 8).

The objective of ensuring a high level of public health protection and the recourse to scientific evidence in taking food safety measures are the threads running through the whole General Food Law. The BSE incident revealed that measures public authorities take in the field of food safety depend on the outcome of scientific risk assessments and that science plays an essential part in achieving a high level of consumer protection.

In this regard, the BSE crisis can be considered the apogee of the transition from modernity to post-modernity. Food in social, political and academic debates has become 'risk' managed by public authorities and society today expects new food products and technologies to have been subject to a risk assessment before being put on the market.

1.7.1 Risk analysis and the precautionary principle

Risk analysis and its three components, risk assessment, risk management and risk communication, although relatively new in public debate, can be traced back to practices known over 5,000 years ago. In Babylon (3,200 BC), a special sect of people was responsible for providing advice in risky, uncertain or important decisions in life. Their tasks would correspond nowadays to risk assessment in the risk analysis terminology (American Chemical Society, 1998: 6).

Well known in the insurance sector, banking, investment and financial market operations, risk analysis refers to strategic management of an organisation and the way various risks are addressed by the organisation's risk management systems.[21] The application of the concept of risk analysis, however, was gradually extending to cover different areas of human activity where potential unwanted consequences had to be assessed, managed and communicated, such as industrial explosions, workplace safety, engineering or the impact of economic developments on ecosystems.[22] With the rise of social regulation, risk analysis was applied to regulatory measures in the areas of the environment and health protection, and finally – towards the end of the 20th century – a paradigm shift in the food safety policy introduced a risk-based approach.

Broadly speaking, what is common for the concept of risk analysis applied to all areas mentioned above can be defined as 'a systematic way of gathering, evaluating and recording information which would lead to recommendations, positions or actions in response to an identified hazard' (WTO, 2000). Risk analysis in social

---

[21] Examples of risks managed by organisations include: financial risks (such as interests rates, foreign exchange, credit), strategic risks (customer changes, competition), operational risks (culture, regulations, supply chain) and hazard risks (natural events, environment, contracts) (AIRMIC *et al.*, 2002: 3).
[22] See also Renn (2006); the Annexes provide a concise overview of different uses of the risk analysis terminology.

regulation, i.e. laws responding to problems affecting the quality of life, is an attempt to link, in a procedural manner, scientific information with policy-making.

Obviously, areas where decision-making depends on science for determining problems and finding responses to them are not confined to food safety. In the EU, the Commission uses different sources of scientific advice to support policies, legislation and regulatory decisions in various areas. Scientific evidence guiding decision-makers is provided by scientific committees grouped around various Directorates; agencies, such as the European Medicines Agency (EMEA) or the European Environmental Agency (EEA); Joint Research Centre (JRC); Scientific and Technical Options Assessment Group (STOA) in the European Parliament; external consultants; or through initiatives aimed at establishing communication platforms, such as the EU Scientific Information Advice in Policy Support (SINAPSE) that promotes information exchange and access to scientific advice produced in Member States.

What truly distinguishes food safety from other areas of risk regulation, however, is that the General Food Law introduces the first comprehensive model of risk analysis applied to the whole regulatory area of food safety at EU and national levels. Risk analysis in the area of food law is a systematic methodology applicable to all *food safety* measures, i.e. measures aimed at the reduction, elimination, or avoidance of a risk to health.[23] The General Food Law sets out the following definitions of the three interconnected components of risk analysis:

- '*risk assessment* means a scientifically based process consisting of four steps: hazard identification, hazard characterisation, exposure assessment and risk characterisation' (Article 3(11) GFL);
- '*risk management* means the process, distinct from risk assessment, of weighing policy alternatives in consultation with interested parties, considering risk assessment and other legitimate factors, and, if need be, selecting appropriate prevention and control options' (Article 3(12) GFL);
- '*risk communication* means the interactive exchange of information and opinions throughout the risk analysis process as regards hazards and risks, risk-related factors and risk perceptions, among risk assessors, risk managers, consumers, feed and food businesses, the academic community and other interested parties, including the explanation of risk assessment findings and the basis of risk management decisions' (Article 3(13) GFL).

The use of scientific advice underpins food safety measures. Risk assessment, however, does not always provide all the insights necessary to formulate a basis for decision makers. In specific circumstances, where the possibility of harmful effects is identified but scientific uncertainty persists, the *precautionary principle* defined in Article 7 GFL is a tool, complementary to the risk analysis principle,

---

[23] Rec. 17 GFL.

enabling decision makers to take measures without having to wait until the reality and seriousness of these risks become fully clarified. Measures based on the precautionary principle are provisional and have to be reviewed within a reasonable period of time, depending on the nature of the risk identified and the type of scientific information needed to clarify the scientific uncertainty.

The assumption that undesirable effects may occur if measures are taken only when scientific data are conclusive underlies the concept of the precautionary principle (Freestone and Hey, 1996). In the cautious social construction of risk, the principle is a mechanism breaking the impasse in the decision-making process in a situation when there are reasonable concerns of adverse effects of a product or technology, but not enough scientific evidence to qualify and compare risks.

Yet, responding to new, complex and variable phenomena, uncertainty is inherent in risk assessment. Risk analysis is applied *because* there is insufficient information, and its role is not only to determine the risk but also to identify uncertainty, understood as errors, information gaps, or insufficient quality of information. As Funtowicz and Ravetz (1993: 742) point out:

> Now that the policy issues of risk and the environment present the most urgent problems for science, uncertainty and quality are moving in from the periphery, one might say the shadows, of scientific methodology, to become the central, integrating concepts. Hitherto they have been kept at the margin of the understanding of science, for laypersons and scientists alike. A new role for scientists will involve the management of these crucial uncertainties; therein lies the task of quality assurance of the scientific information provided for policy decisions.

Uncertainty, often accompanied by scientific controversy, is difficult to address in policy-making. Risks are both social constructions and physical phenomena (OECD, 2003). Although it is crucial to evaluate them as objectively as possible, management decisions must be tuned across a multitude of dimensions of risk, taking into account not only facts, but also values and uncertainties. In decisions based on the precautionary principle, risk management takes a standpoint towards these uncertainties: priority is given to the protection of human health. The principle is thus not neutral towards uncertainty – it is biased in favour of safety (Bodansky, 1994), and, as such, closely linked to the obligation to assure a high level of health protection in the pursuit of EU policies stated in Article 114(3) of the Treaty on the Functioning of the European Union (TFEU – for the text of Article 114 TFEU see Appendix 1).

1.7.2 Other legitimate factors in risk analysis

To determine the permitted level of a contaminant or to authorise a novel food or a new additive, risk assessment is carried out by scientific experts. In food safety regulation science sets everything in motion, but in most cases, scientific evaluation alone is not sufficient. A number of other pertinent considerations have to be legitimately taken into account in risk management (Figure 1.3).

The General Food Law mentions societal, economic, traditional, ethical and environmental factors, as well as the feasibility of controls, as examples of other legitimate factors.[24] Hence, risk management includes balancing the conclusions of the scientific evaluation of risk against social concerns, the benefits that a product or technology brings and economic implications. How to reconcile in a consistent and democratic way these – often conflicting – needs of technical expertise and public acceptance remains one of the fundamental issues of risk regulation (Fisher, 2007).

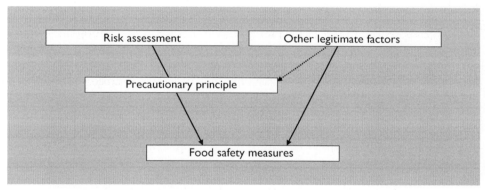

Figure 1.3. Risk analysis methodology.

1.8 Risk analysis and food trade

To protect human health, food safety legislation intervenes in market mechanisms by restricting or prohibiting the use or sale of certain products. Carrots, in which the level of lead does not exceed 0.10 mg/kg, are considered safe for human health. If the level of lead exceeds 0.10 mg/kg, they become *res extra commercium*. Trade rules are thus the opposing dynamics to risk regulation, within the EU internal market as well as in the international arena. The acceptable level of risk is determined on the basis of risk assessment provided by independent experts. From a trade law perspective, science plays the role of an objective 'arbiter' judging whether a

---

[24] Rec. 19 GFL.

measure is justified, i.e. whether it serves the purpose of protecting human health and is not used as an excuse for protecting domestic products.

Article 36 TFEU allows exceptions by Member States to the free movement of goods within the EU on grounds, i.e. of the protection of health and life of humans, animals or plants. The Court of Justice of the European Union construes these exceptions narrowly. National measures aimed at the protection of human health which restrict the free movement of goods between Member States must be confined to what is actually necessary to ensure the safeguarding of public health and must be proportional to the objective pursued, which could not have been attained by measures which are less trade restrictive.[25]

Proportionality weighs on the one hand the magnitude of risks to human health, and on the other hand the economic consequences of a food safety measure (Seiler, 2002). Recourse to science filters out unjustified barriers to the free movement of foodstuffs, requiring scientific proof for food safety measures and excluding all measures going beyond what is necessary to achieve the desired level of health protection.

The grounds set out in the TFEU concerning trade-restrictive national measures aimed at the protection of human life and health resemble those found in the World Trade Organisation (WTO) regime. The system of trade rules aims to limit regulatory discretion as much as possible, by making 'scientific evidence' the criterion deciding which regulations are justified. The WTO Agreement on Sanitary and Phytosanitary Measures (SPS) requires that national SPS measures be based on risk assessment. Article 2(2) SPS states that WTO members shall ensure that any sanitary or phytosanitary measure is 'based on scientific principles and not maintained without sufficient scientific evidence'. Article 5 SPS introduces risk assessment as a method for determining the appropriate level of health protection, carried out according to the internationally developed techniques.

The SPS Agreement does not contain any food safety standards – these are developed outside the WTO.[26] The Agreement is confined to a purely procedural requirement that sanitary or phytosanitary measures be based either on international standards or on risk assessment. It sets out a meta-framework of requirements for national (and regional – EU) measures.

---

[25] Case 192/01, *Commission* v. *Denmark*, [2003] ECR I-9693, para. 45.
[26] Of particular importance for food safety are standards developed by the Codex Alimentarius Commission, established by the Food and Agriculture Organization (FAO) and the World Health Organization (WHO). The Codex Alimentarius Commission develops food standards, guidelines and other texts (e.g. codes of practice) with main aims to protect human health and to ensure fair practices in international food trade. Codex standards serve as a reference point for the WTO law – national measures based on Codex standards enjoy a presumption of compliance with the SPS Agreement.

1.9 Research objective and questions

In the era of risk society, regulatory activities of states are predominantly characterised by a focus on risk: food safety, health protection, regulation of chemicals (EU REACH), product safety, vehicle safety, and environmental governance all are examples of settings where risks are dealt with. Risk regulation is becoming one of the dominant forms of regulation in the EU internal market.

In risk governance, decision-makers rely heavily on technical and scientific expertise. Science draws a demarcation line between the objectives of the effective functioning of the internal market and the assurance of a high level of protection of human health in food regulation. Establishing policy in a democratic manner, however, requires taking into account a broad array of societal values and preferences, such as traditional, ethical, and societal concerns.

Because of the special meaning of food and historical context, the marriage between scientific expertise and public opinion is particularly troubled in EU food safety regulation, referred to by Ansell and Vogel as 'contested governance' (Ansell and Vogel, 2006a).[27] For this reason alone, food safety regulation provides an excellent lens into developments in contemporary law-making processes. Relating to the role of science in policy making, these technocratic styles of policy making, including multi-level regulation: international, EU and national, may pose accountability problems to democracies.

Food safety so far is also the only area of risk regulation where a comprehensive risk analysis model and the precautionary principle have been introduced as general principles governing the food safety policy, applied to both EU and national measures. This prescribed science-based character of food safety measures is set out not only at EU, but also at international level by the WTO regime. Hence, food law captures legislative dynamics and interrelations in a field of regulation based on the risk analysis methodology in the increasingly integrated system of multilevel governance, including the international level.

For all these reasons, the objective of this book is to analyse:

> the concepts of risk analysis and the precautionary principle established in the General Food Law and their impact on food safety regulation in EU multi-level governance.

---

[27] 'Contested governance' is understood as a 'particularly intense and broad-based conflict about the foundational assumptions and institutional frameworks through which a policy domain is governed' (Ansell and Vogel, 2006a).

On the basis of the research objective the following questions are formulated:

- What is the constitutional framework for risk analysis?

  Which Treaty provisions relate to food law? What is the division of competences between the EU and Member States in this field? What is the case-law of the European Courts dealing with scientific evidence and the precautionary principle applied to both European and national food safety measures in this context? What is the impact of the science-based food governance introduced by the General Food Law on the functioning of the EU internal market?

- Do the risk analysis methodology and the precautionary principle apply only to implementing regulatory measures or do they set limitations on the legislator as well?

  Does the requirement that food safety measures be science-based exclusively apply to authorization decisions and other administrative activities 'under legislation' or does it also have impact on legislation itself? To what extent does the risk analysis methodology set limitations on the EU legislator in deciding on food safety measures that restrict trade? Can EU food safety measures be tested under international trade obligations? If so, what is the outcome? Does EU food safety legislation comply with the general principles of EU food law?

- What is the scope of application of the precautionary principle as defined in Article 7 of the General Food Law?

  How did the concept of the precautionary principle evolve from a general political guidance into a legally defined principle of food law? What are the main constituents of the precautionary principle defined in Article 7 GFL and how does the risk analysis methodology further narrow the scope of the principle? Do existing 'precautionary' policy orientations and the concept of the precautionary principle shaped by case-law prior to the General Food Law fit within the concept codified in the definition? Does the new definition require rethinking of the character of 'precautionary' measures that can actually be based on the precautionary principle?

- How are other legitimate factors incorporated into the decision-making process at EU and national levels of food safety governance?

  The role of factors other than science in food law making is not clearly defined – do the Treaty provisions relating to the free movement of goods shed light on their role in EU food safety governance? The extent to which risk managers can deviate from scientific evaluations in considering these factors depends on how much discretion is conferred on public authorities. Does this discretion apply in the same way at EU and national level? Does the European judiciary prefer scientific opinions of EU advisory bodies and EFSA to national risk assessments? How do internal market mechanisms react if the inclusion of factors other than science results in a measure more restrictive than recommended in risk assessment? What happens if a measure ensures a level of protection that is lower than advocated by scientific opinion?

The four main questions are addressed in four chapters, based on articles published in scientific journals or book chapters. The chapters focus on crucial aspects of the application of the risk analysis methodology and the precautionary principle, which have also been identified as major literature gaps. The approach applied in this research breaks new grounds in the field of EU food law by (re-)interpreting the concepts of risk analysis and the precautionary principle introduced by the General Food Law under the assumption that this comprehensive food safety policy sets out general principles for making food laws in the European Union.

The principles set out in the General Food Law do not operate in a regulatory vacuum. Although the term 'food safety' does not appear in the Treaty itself, in so far as it aims at the protection of human health, food law is almost entirely centralised at EU level. Food law remains a cross-sectoral matter whose legal bases are predominantly the EU internal market (Article 114 TFEU), and, to a certain extent, agriculture (Article 43 TFEU). Food safety policy areas are thus shared competences between the EU and the Member States (Article 4 TFEU). The multi-level food safety regulation and the general principle of food law are thus viewed through the lens of Article 114 TFEU and its opt-out (safeguard) clauses for Member States.

Therefore, although the way in which different tasks are fulfilled largely depends on the institutional structure of the EU (Kapteyn and VerLoren van Themaat, 1998),[28] this study does not include the myriad of political factors that shape food law, but focuses instead on the analysis of substantive provisions of European Union food law, consisting in interpretation and systematisation of legal norms and formulating conclusions about their application.[29]

In doing so, classical methods of interpretation of law are combined: literal, contextual, and teleological. When a provision is sufficiently clear and unambiguous, literal interpretation usually provides a satisfactory basis for analysis. Such interpretation, however, has to fit in the design of the Treaty and general and specific objectives of EU law, as well as Court of Justice case-law and international law. Finally, the challenge of this study also consists in dressing risk analysis terminology – used predominantly in the realm of toxicology, microbiology or chemistry – in legal clothing.

---

[28] In reality, the interaction between substantive and institutional aspects of food safety policy determines the decision-making process. The European Parliament and the Council, acting in accordance with the ordinary legislative procedure, often considerably dilute scientific inputs on which the Commission's proposals are based. E.g. in 2010 the European Parliament blocked the Commission's proposal for adding bovine and porcine thrombin as food additives. These enzymes are used as 'meat glue' – they can bind separate pieces of meat together and create a product of a desirable form. Despite EFSA's positive risk assessment of 'meat glue', the Parliament estimated that this method could be misleading and that there was no clear benefit for consumers. The proposal was rejected. See Chapter 5 for a legal analysis of the consequences of the inclusion of other legitimate factors resulting in more restrictive measures.

[29] This type of legal research is called 'legal dogmatics' or 'analytical study of law' (Peczenik, 2009).

2. From mutual recognition to mutual scientific opinion?*

Constitutional framework for risk analysis in EU food safety law

Abstract

Regulation 178/2002 (the so-called General Food Law – GFL) codifies risk analysis as the core principle of the modern food safety policy. This chapter places the GFL in EU multi-level food safety governance and analyses the impact of risk analysis, the precautionary principle and mechanisms of scientific governance introduced by the GFL on both national and EU legislation. It discusses the case law of the European Courts dealing with scientific evidence and the precautionary principle applied to both European and national food safety measures. The chapter concludes with some observations on the role of the risk analysis methodology in the Union internal market.

2.1 Introduction

Food safety remains a contentious issue in the European Union, where the requirement of assuring a high level of protection of human health has to be reconciled with the principle of the free movement of goods within the internal market. Balancing these two imperatives creates tensions, especially as policy responsibility is distributed among Union and national levels of governance.

The European Commission recognized the specificity of the food sector already in 1985, stating in its Communication on the Community legislation on foodstuffs that the creation of a common market for foodstuffs would be especially challenging due to 'the extreme sensitivity of public opinion in this field', and the abundance of national laws parallel to the almost complete lack of Community standards (EC, 1985b).

The subsequent stages of the development of a common market substantiated that supposition. Despite intensive efforts to harmonise national laws, the food sector remained subject to numerous trade barriers. The latest regulatory reform in the area of food safety, crowned by Regulation 178/2002 laying down the general principles and requirements of food law, establishing the European Food Safety

---

* Reprinted from Food Policy, 34, Szajkowska, A., From mutual recognition to mutual scientific opinion? Constitutional framework for risk analysis in EU food safety law, 529-538, 2009, with permission from Elsevier.

Authority and laying down procedures in matters of food safety (the General Food Law – GFL)[1] is a step further than harmonization. Some authors see it as the development of a new EU policy (Ugland and Veggeland, 2006).

As one of the principles underlying the new approach, the General Food Law recognises – for the first time in EU law – a comprehensive risk analysis model to be followed by the EU and Member States in establishing food safety legislation.[2] Risk analysis, consisting of three interconnected components – scientific risk assessment, risk management and risk communication – is a methodology used in risk regulation to provide consistency in decision making in areas where uncertainty is an inherent feature (Majone, 2003). Food safety law is a typical example of risk regulation – defined by Hood *et al.* (2001: 4) as a 'governmental interference with market or social processes to control potential adverse consequences to health'.

This chapter analyses the impact of the principle of risk analysis, the scientific institutional framework and procedures established in the General Food Law on the functioning of the EU internal market. It starts with a short description of the genesis, scope of application and content of the General Food Law, based on an assumption that the Regulation creates a new integrated EU policy. We then discuss the elements of this policy that are related to scientific governance: the principle of risk analysis, the new institutional framework and procedures. In the next section, we place the General Food Law in multi-level food safety governance in the EU, and we look at the division of competences between the EU and Member States and legislative dynamics in this field. Then, we analyse the case law of the European Courts dealing with scientific evidence and the precautionary principle applied to both European and national food safety measures. In the case of national food safety measures we take into account all possibilities under the EC Treaty where Member States can either restrict the free movement of goods on grounds of the protection of human health in the non-harmonised area, or derogate from EU-harmonized legislation on the basis of Article 114(4) TFEU or safeguard clauses. The precautionary principle is discussed separately in the last section. The chapter concludes with some observations on the role of the risk analysis methodology in the EU internal market.

2.2 General Food Law – the new EU policy

2.2.I Genesis

For a long time European food law was devoted to the completion of the internal market and its development was predominantly a 'spillover' of that main goal (Vos, 1999: 9). The situation changed considerably after the BSE crisis of the mid-nineties,

---

[1] OJ 2002, L 31/1.
[2] Art. 6 GFL.

when the British government admitted publicly that a connection between the BSE and the Creutzfeldt-Jakob disease could not be excluded. The 1997 European Parliament enquiry report revealed serious shortcomings in the handling of the crisis at EU level, and accused the Commission of protecting the market and business interests over those of the consumer (European Parliament, 1997).

These shortcomings demonstrated the need of a new, integrated approach to food safety at EU level. The report was shortly followed by a series of EU actions designed to improve the policy in this field and to regain citizens' trust in the EU institutions. The food safety reform was outlined in the Commission Communication on Consumer Health and Food Safety (EC, 1997a), Green Paper on the General Principles of Food Law in the European Union (EC, 1997b) and — a little over two years later — in the White Paper on Food Safety aimed at the creation of a 'comprehensive and integrated approach to food safety' (EC, 2000a). The new system of principles and requirements of food law at EU level, together with new procedures in matters of food safety and a new institution – the European Food Safety Authority (EFSA) – have been enshrined in the 2002 General Food Law.

2.2.2 Scope of application

The General Food Law lists as many as four Articles of the TFEU as its basis: Article 43 for harmonization in agricultural sector, Article 114 for approximation of laws for the purpose of the internal market, Article 168(4)(b) for measures in the veterinary and phytosanitary fields having as their direct objective the protection of public health, and Article 207 referring to the common commercial policy. This fact, unusual for Community legislation, already illustrates how broad and complex the scope of the Regulation is (Costato, 2003).

The general principles of European food law apply to 'all stages of production, processing and distribution of food, and also of feed produced for, or fed to, food-producing animals'[3] (the *farm to fork* continuum – only primary production for private domestic use or the domestic preparation, handling or storage of food for private domestic consumption are exempted from the scope of application of the GFL).[4] 'Food law' is defined in the Regulation as 'laws, regulations and administrative provisions, governing food in general, and food safety in particular, whether at Community or national level'.[5] Thus, the Regulation sets out general principles of food safety for all legislation in this field, at all levels of European governance. All fundamental aspects of food safety are harmonised at EU level, regardless of whether their scope is limited to local trade or reaches beyond the borders of one Member State. The Regulation provides, i.e. an EU definition of

---

[3] Art. 4(1) GFL.
[4] Art. 1(3) GFL.
[5] Art. 3(1) GFL.

'food',[6] it establishes a traceability system,[7] Rapid Alert System for Food and Feed (RASFF),[8] and emergency procedures,[9] it sets out an overarching requirement that food shall not be placed on the market if it is unsafe,[10] and defines public and private responsibilities for assuring food safety.[11] This comprehensive approach to food safety is often characterized as the emergence of a new integrated Union policy (Lafond, 2001: 17; Ugland and Veggeland, 2006). Ugland and Veggeland (2006) distinguish two ways of integrating the new food safety policy: directly – by laying down general principles, and indirectly – by creating an institutional framework for better scientific governance in this field. These elements will be discussed in the next two sections.

2.2.3 Principle of risk analysis: international context

Among the general principles enshrined in the General Food Law an interesting novelty is risk analysis and the precautionary principle, which have been codified for the first time at EU level.

The inclusion of the principle of risk analysis reflects the EU's international obligations under the Agreement on the Application of Sanitary and Phytosanitary Measures (SPS Agreement) concluded within the framework of the World Trade Organisation (WTO). Because the EU is obliged to provide scientific justification for its sanitary and phytosanitary measures at international level, the scientific regime established by the SPS Agreement has penetrated through WTO-friendly legislation into the EU system.[12] The SPS Agreement is an example of a methodological regulation: it imposes procedures to be followed by WTO members in drawing up sanitary or phytosanitary legislation that affects international trade. The Agreement tends to limit the regulatory discretion to a minimum:[13] measures must be based on appropriate risk assessment and cannot be maintained without sufficient scientific evidence. The only exception is provided in Article 5.7 of the SPS Agreement, which allows WTO members to adopt *provisionally* sanitary or phytosanitary measures when relevant scientific evidence is insufficient. This provision is generally seen as the reflection of the precautionary principle, although the SPS Agreement does not refer to this principle explicitly.[14]

---

[6] Art. 2 GFL.
[7] Art. 18 GFL.
[8] Art. 50 GFL.
[9] Art. 53 and 54 GFL.
[10] Art. 14(1) GFL.
[11] Art. 17 GFL.
[12] Art. 13 GFL. On the influence of WTO regime on EU law see Alemanno (2007) and Skogstad (2001).
[13] On limitations set by the SPS Agreement on national discretionary powers see Slotboom (1999).
[14] See Appellate Body Report. 1998. *EC – Measures Concerning Meat and Meat Products*, WT/DS26/AB/R; WT/DS48/AB/R, adopted on 13 Feb. 1998.

Risk is a function of the probability of an adverse health effect and the severity of that effect, consequential to a hazard.[15] The purpose of risk assessment is to evaluate the risk. This scientific evaluation consists of four steps: hazard identification, hazard characterization, exposure assessment and risk characterization.[16] On the basis of scientific risk assessment, decision makers define political objectives to determine the level of risk acceptable for the society. The WTO regime allows its members to determine the level of protection of human life and health that they deem appropriate.

Whereas the SPS Agreement tends to limit the regulatory discretion, the intention of the European legislator seems to be the opposite. Article 6(3) GFL stipulates that risk management shall take into account not only the results of risk assessment, but also the precautionary principle and 'other factors legitimate to the matter under consideration'. The precautionary principle – now codified in Article 7 GFL – allows decision makers to act without having to wait until the reality and seriousness of risk to health are fully demonstrated, if the available information identifies possible harmful effects on human health of a product or activity. Other legitimate factors include societal, economic, traditional, ethical and environmental factors and the feasibility of controls.[17] Their inclusion in the GFL risk analysis model raises an interesting question about the extent to which risk managers can base their decisions on factors other than science to remain in compliance with the supranational risk regulation regime established by the WTO.

The General Food Law stipulates in Article 6(1) that 'food law shall be based on risk analysis except where this is not appropriate to the circumstances or the nature of the measure'. Risk analysis is a methodology applied to decisions in the field of food safety, i.e. the part of food law aiming at the reduction, elimination or avoidance of a risk to health.[18] Certainly, not all food law requires recourse to scientific risk evaluation. For example, many technical food standards that do not have as their objective the protection of the health and life of humans, or legislation relating to consumer information, are excluded from the scope of risk analysis.[19]

2.2.4 The European Food Safety Authority and new procedures

The increasing harmonisation of food safety legislation at EU level and the introduction of the risk analysis methodology have resulted in the creation of a new institutional framework. The European Food Safety Authority was designed not only to reinforce scientific support for EU legislation, but also to help regain

---

[15] Art. 3(9) GFL.

[16] Art. 3(11) GFL.

[17] Rec. 19 GFL.

[18] Rec. 17 GFL.

[19] However, consumer information related to the protection of human health, such as, e.g. nutrition and health claims, is based on risk analysis.

consumer trust in EU institutions, shattered by the BSE crisis. The European legislator based its policy on risk analysis and expert involvement in the law-making process, but did not delegate regulatory powers to an independent technocratic agency.[20] Instead, it regrouped existing scientific committees under a separate institution, independent from risk managers and transparent, and as such more 'visible' to consumers. Yet, given that the risk analysis principle explicitly introduces a politics of knowledge in the field of food safety, food safety regulation in the EU will largely depend on EFSA's say (Chalmers, 2003).

In EU multi-level risk governance EFSA collaborates in a network with institutions carrying out similar tasks in Member States.[21] The EU did not empower EFSA with a decisive voice in case of diverging scientific opinions.[22] In practice, such ultimate voice of EFSA, deciding whether risk assessment for a national food safety measure is convincing and scientifically sound, would turn EFSA into a European arbiter. Instead, in order to avoid inconsistency between national and EU approach and – in consequence – possible barriers to EU trade, the GFL introduces a new procedure to deal with conflicting scientific opinions. According to Article 30 GFL, EFSA is obliged to be vigilant to identify – at an early stage – any diverging scientific opinions between the Authority and a national or EU body, and to cooperate with a view to either resolving the divergence or preparing a joint document explaining the divergent opinions, which shall be made public.

Another new procedure introduced by the General Food Law is mediation set out in Article 60 GFL, where EFSA also plays an important role. The mediation procedure involves a situation where a Member State considers that a measure taken by another Member State is either incompatible with the General Food Law or likely to affect the functioning of the internal market. In this case, the Member State refers the matter to the Commission, which – if agreement cannot be reached – may request EFSA to give an opinion on the relevant contentious scientific issues. The mediation procedure is especially meant to provide the parties concerned with the opportunity to discuss scientific issues relating to the measure in question and to reconsider them in the light of the scientific opinion provided by EFSA, before the matter is brought to the attention of the Standing

---

[20] On the concept of European agencies and the delegation of powers see Chiti (2000, 2002), Dehousse (1997), Kreher (1996), Kreher and Martines (1996), Lafond (2001) and Vos (1999, 2000b).
[21] Reg. 2230/2004 laying down detailed rules for the implementation of Reg. 178/2002 with regard to the network of organisations operating in the fields within EFSA's mission, OJ 2004, L 379/64. On national agencies see, e.g. on the Spanish Food Safety Agency: Franch Saguer (2002). In Italy, transitional arrangements have been made on the way to the creation of an Italian Food Safety Agency – in 2004 the National Committee for Food Safety was set up – see Cassese (2002). It must be noted, however, that Member States are not obliged to create national independent food safety agencies. In Poland, risk assessment in the field of food safety is carried out by 7 national research institutes.
[22] For a critical opinion on this solution see Mandeville (2003).

Committee on the Food Chain and Animal Health (SCFCAH) under comitology procedures.[23]

In the food safety legislation harmonised at EU level, risk assessment is centralised and entrusted to EFSA solely. An example can be found in the Regulation on genetically modified food and feed,[24] where only EFSA issues scientific opinions concerning GM products and the role of national authorities can be compared to the one of a box office (Van der Meulen and Van der Velde, 2008: 302).[25] Similarly, EFSA evaluates health claims on foods, as well as food additives, food enzymes and food flavourings before they are authorized for use.[26] Subsequent revisions of legislation existing prior to the GFL are replacing existing authorisation frameworks for one centralized procedure with EFSA as the sole risk assessor.[27]

2.3 EU multilevel food safety governance

2.3.1 EU and national competences in the field of food safety: place of the GFL in the Community legal system

The EU does not have exclusive competence in the field of food safety, but, since this area is closely related to the internal market, national competences can be pre-empted through total harmonisation via Article 114 TFEU. It means that once total harmonization has taken place, Member States are not allowed to introduce national measures on grounds of the protection of health and life of humans, animals or plants (sanitary and phytosanitary measures) or to invoke other possible exceptions listed in Article 36 TFEU,[28] justifying prohibition or restriction on the free movement of goods within the Community. The possibility for Member States to derogate from harmonisation measures is limited to exceptions provided in Article 114 TFEU. This is in conformity with the *Tedeschi* case-law: once total harmonisation has taken place, Article 36 TFEU does not apply.[29]

---

[23] In this context, it is also interesting to note that – according to Art. 59 GFL – the SCFCAH has a broad competence to examine issues falling outside normal comitology procedure (Vos, 2005). The SCFCAH can examine 'any issue falling under those provisions, either at the initiative of the Chairman or at the written request of one of its members'.

[24] Reg. 1829/2003 on genetically modified food and feed, OJ 2003, L 268/1.

[25] EFSA can, however, engage a national assessment body in the execution of its tasks.

[26] Reg. 1924/2006 on nutrition and health claims made on foods, OJ 2007, L 12/3; Reg. 1331/2008 establishing a common authorization procedure for food additives, food enzymes and food flavourings, OJ 2008, L 354/1.

[27] See, e.g. Proposal for a revised Novel Foods Regulation, COM(2007) 872 final.

[28] The other exceptions are: public morality, public policy and public security, the protection of national treasures possessing artistic, historic or archaeological value, the protection of industrial and commercial property.

[29] See Case 5/77, *Tedeschi* v. *Denkavit*, [1977] ECR-1555.

Article 114 TFEU envisages three possibilities to derogate from harmonised European legislation.[30] Two of them are relevant to the field of food safety. The first possibility concerns the right of a Member State to *maintain* national measures (existing before the harmonisation of laws at EU level) on the grounds of the major needs listed in Article 36 TFEU, or relating to the protection of the environment or the working environment. The second possibility to deviate from harmonisation measures is a clause, introduced by the Treaty of Amsterdam in Article 114(5), allowing Member States to adopt *new* national legislation, aiming at the protection of the environment or the working environment. In this case, the reference to Article 36 TFEU does not exist, which means that the introduction of *new* national measures derogating from harmonized EU legislation in the area of food safety is not possible. The third possibility provides for the inclusion of 'safeguard clauses' in harmonized legislation, authorizing Member States to take, for one or more non-economic reasons referred to in Article 36 TFEU, provisional measures, which are subject to an EU control procedure (Article 114(10) TFEU).[31]

In the *Cindu Chemicals* judgment the European Court of Justice (ECJ) favors teleological interpretation, according to which legislation based on Article 114 TFEU should generally be interpreted as having an exhaustive character. Because the rationale of measures based on this Article is to eliminate obstacles to trade within the internal market, it would not be possible to attain this objective if the Member States were free to maintain or adopt national measures on other grounds than derogations under Article 114 TFEU.[32]

What type of harmonization does the General Food Law constitute? At first glance, one may wonder if a comprehensive food safety policy of such a broad scope established at EU level leaves any competences at all for the legislator at national level. The new food safety policy creates 'a general framework of a horizontal nature',[33] integrating existing and future national and EU legislation under common principles. The legislation existing before the General Food Law had to be adapted in order to comply with the principles set out in Articles 5-10 GFL by 1 January 2007.[34] In the non-harmonised area, the Regulation does not prevent Member States from invoking Article 36 TFEU as a basis of their national measures aimed at the protection of human life and health in the absence of specific EU legislation. Specific food safety provisions at EU level are adopted in accordance with Article

---

[30] On the derogations under Art. 114 TFEU see De Sadeleer (2003).

[31] It also should be noted that harmonisation of the production and marketing of agricultural products listed in Annex I to the Treaty is based on Art. 43 TFEU, which does not provide for any national derogations.

[32] Joint Cases C-281/03 and C-282/03, *Cindu Chemicals BV and Others*, [2005] ECR I-8069.

[33] Art. 4(1) GFL.

[34] 'General principles' and 'general requirements' have a similar meaning – in the initial Commission proposal they were included in one section called 'Principles and requirements of food law'. The requirements listed in Articles 11-20 apply from 1 January 2005 (except for Art. 13 referring to the development and use of international standards, which entered into force with the Regulation).

114 TFEU when harmonisation of national provisions is necessary for the proper functioning of the internal market. This is explicitly stated in Article 14(7) and 14(9) GFL and confirmed in recent case law, where the ECJ held that, according to Article 14(9) of Regulation 178/2002, in the absence of specific EU rules laid down in those provisions, national rules may be applied without prejudice to the provisions of the Treaty.[35] Thus, the General Food Law, by laying down general principles governing food and feed, constitutes only a first stage of harmonisation.

2.3.2 National laws: free movement of goods and mutual recognition

The General Food Law is laconic about the non-harmonised provisions of food safety law. Article 14(9) GFL reads:

> Where there are no specific Community provisions, food shall be deemed to be safe when it conforms to the specific provisions of national food law of the Member State in whose territory the food is marketed, such provisions being drawn up and applied without prejudice to the Treaty, in particular Articles 28 and 30 thereof [34 and 36 TFEU].

Because food safety is closely related to the functioning of the internal market, competences in this area are delineated by the Treaty provisions referring to the free movement of goods. Article 34 TFEU sets out the general rule for the free movement of goods: it prohibits quantitative restrictions on imports and all measures having equivalent effect between Member States. In *Dassonville*[36] the ECJ clarified that 'measures having equivalent effect' can be not only measures applied to imported goods solely, but also those giving equal treatment to domestic and imported production. Thus, every national standard can constitute a potential measure having an effect equivalent to quantitative restriction on imports. Possible exceptions to the free movement of goods are listed in Article 36 TFEU or can be derived from the so-called 'rule of reason'.[37]

---

[35] Case C-319/05, *Commission* v. *Germany*, [2007] ECR I-9811, at 84.
[36] Case 8/74, *Procureur du Roi* v. *Dassonville*, [1974] ECR 837.
[37] The exceptions to the free movement of goods set out in Art. 36 TFEU have to be interpreted strictly, but the rule of reason, developed by case law to counterbalance the broad interpretation of the principle of the free movement of goods, shows that other grounds, not explicitly mentioned in the TFEU, may also constitute justified barriers to trade. The rule of reason justifies national measures hindering national trade which are necessary to assure fair trade, to protect consumers, and interests and values that are not contained in Art. 36 TFEU, but are recognized by the Court as important. Examples of the rule of reason justifications include effectiveness of fiscal supervision, fairness of commercial transactions or the protection of the environment. Some of the interests previously recognized by the rule of reason are now included in the EU area of harmonisation under Art. 114(3) TFEU.
It must also be noted that, by their very nature, measures aimed at the protection of human life and health, as well as the other exceptions listed in Art. 36 TFEU, may legitimately apply to imported goods solely, whereas food law measures relating to consumer protection (falling under the 'rule of reason') must treat domestic and imported goods in the same way (Kapteyn and VerLoren van Themaat, 1998: 675).

The original approach to eliminate obstacles to the free movement of goods between the Member States consisted of the total harmonisation of laws. Early food law was generally characterised by detailed provisions unifying technical (compositional) standards at EU level. This task, consisting of issuing recipe laws for different foodstuffs (the so-called 'vertical' harmonization), required approximation of national laws regulating the market on an unprecedented scale. Undoubtedly, this approach was doomed to failure in the long run.

The Court remedied the situation with the principle of mutual recognition, derived from Article 34 TFEU and coined in the celebrated *Cassis de Dijon* judgment.[38] The principle of mutual recognition generally precludes a Member State from prohibiting the sale of a product which does not conform to domestic regulations, but which has been lawfully marketed in another Member State, unless such prohibition can be based on Article 36 TFEU or the rule of reason. In other words, the requirement of the free movement of goods has been recognized as a 'meta-norm' in the case of conflict of national laws (Joerges, 2006: 4). The principle could be then substituted for the previous vertical harmonization approach. In the absence of harmonized EU legislation, Member States can lay down rules containing specific requirements for products. However, in order to ensure the free movement of goods within the European Union, they are obliged to accept in their territories products complying with the regulations of another Member State, even if technical or qualitative characteristics of these products differ from domestic regulations.[39]

2.3.3 New approach to harmonisation

The consequence of the *Cassis de Dijon* ruling is that it is no longer necessary to harmonize painstakingly all compositional requirements for foodstuffs in EU legislation (EC, 1985b: 12). The free movement of goods, as far as it does not affect the national level of food safety, can – in principle – be ensured by the principle of mutual recognition.[40]

---

[38] Case 120/78, *Rewe-Zentral AG* v. *Bundesmonopolverwaltung für Branntwein* (more popularly known as the *Cassis de Dijon* judgement), [1979] ECR 649.

[39] See also Reg. 764/2008 laying down procedures relating to the application of certain national technical rules to products lawfully marketed in another Member State, OJ 2008, L 218/21. The Regulation, i.e. puts the burden of proof on a Member State, by setting out some procedural requirements for the competent authorities intending to take restrictive measures about the product.

[40] The Commission justified this approach as follows (EC, 1985b: 9):
'it is neither possible nor desirable to confine in a legislative straitjacket the culinary riches of ten (twelve) European countries;
legislative rigidity concerning product composition prevents the development of new products and is therefore an obstacle to innovation and commercial flexibility;
the tastes and preferences of consumers should not be a matter for regulation'.

However, in the area of food safety, where the main purpose of legislation is the protection of public health, the application of mutual recognition is hindered because Member States can invoke Article 36 TFEU as a basis for their legislation and thus claim exemption from including the mutual recognition clause in their measures. As a consequence, in most cases – with the exception of national technical standards not directly related to the protection of human life of health where the internal market objective can be attained by mutual recognition – harmonization at EU level remains the only way to assure the free movement of foodstuffs.

For this reason, since 1985 (when the objective of completing the internal market was set out), the European Union, in principle, has tried to confine the harmonization of food law to provisions aiming at the protection of human, animal or plant health. Its activity in the foodstuffs sector has been concentrated on issuing 'horizontal' directives of a broad application, concerning the use of additives, materials and articles in contact with food, irradiation of foodstuffs, etc. (EC, 1985a: 21). According to the 'New Approach' to harmonisation and standardisation, set out in the 1985 Council Resolution, EU legislation in the field of technical regulations has been limited to the essential requirements needed to ensure the free movement of goods throughout the EU, while the task of drawing up technical specifications for the production and placement on the market is entrusted to independent standards organizations (European Council, 1985). Their standards are not mandatory, but at the same time products manufactured in conformity with them are presumed to comply with the essential requirements established by EU measures.

The new strategy is in line with the principle of subsidiarity. Accordingly, first, the choice between harmonization and the principle of mutual recognition has to be made, and then, if need be, an appropriate method of harmonization has to be framed, limiting legislative intervention at EU level to a necessary minimum (EC, 1993).[41] The New Approach should not however be interpreted as 'minimum rules', but rather as 'necessary rules' (Grey, 1993). Actually, what happened in Europe after the programme on the 'Completion of the Internal Market' was launched did not result in the loosening of national food safety requirements in order to attract businesses and enhance the competitiveness of Member States' economies ('race

---

[41] Most of the EU recipe-laws (the 'early' food law) examined in the light of subsidiarity requirements, raised concerns about their justifiability. The Edinburgh European Council in 1992 recommended that, under the New Approach, these detailed technical specifications ('vertical' directives) be simplified and replaced by minimum requirements for products circulating freely within the EU market (European Council, 1992). Hence, although there still remain quite a few recipe-laws relating to food products at European level, e.g. legislation setting standard for honey, fruit juices, jams, jellies, marmalade, coffee, natural mineral waters, minced meat, eggs, not to mention chocolate – the first, and probably the most famous 'Europroduct', the detailed technical regulations have been gradually replaced and amended in accordance with subsidiarity requirements. See, e.g. in reference to cocoa and chocolate standards Dir. 73/241/EEC (OJ 1973, L 228/23), replaced by Dir. 2000/36/EC (OJ 2000, L 197/19).

to the bottom'), but rather in strengthening cooperation and activity in the area of risk regulation at EU level (Joerges, 2006; Vos, 1999: 131-134).

2.3.4 Centralisation of EU food safety law

Where full harmonisation measures referring to the protection of public health exist at EU level, recourse to Article 36 TFEU is no longer possible. In this situation, the European Union is considered to 'have taken over' the area of regulation. In harmonized fields national measures are permitted only if the harmonization measure allows them explicitly through safeguard clauses or if some aspects have explicitly been left for the national regulations of the Member States – the latter can be found in the Hygiene Regulations for example.[42] The inclusion of safeguard clauses is virtually the case for all European legislation relating to foodstuffs (Berends and Carreño, 2005).

Today, the majority of food safety legislation is harmonized at the European level and the importance of Article 36 TFEU and the rule of reason in relation to foodstuffs is gradually decreasing.

2.4 The principle of risk analysis and national food laws

2.4.1 Case law on Article 34 TFEU: pre-market approvals and burden of proof

In the non-harmonised area of food law, the intra-EU trade in foodstuffs can be affected by the right of a destination Member State to set a level of health protection that it considers appropriate and to verify whether a given product complies with this level.[43] The EU Courts, in their abundant case law relating to restrictions on the free movement of goods on grounds of the protection of health and life of humans, animals or plants, had elaborated a standard of review based on scientific evidence before the principle of risk analysis was explicitly worded into EU food safety legislation. An overview of case law relating to the interpretation of Articles 34-36 TFEU gives an interesting insight into the risk analysis methodology.

---

[42] Art. 13(3) Reg. 852/2004 on the hygiene of foodstuffs, OJ 2004, L 226/3.
[43] Case 104/75, *De Peijper*, [1976] ECR 613, at 15, Case 192/01, *Commission* v. *Denmark*, [2003] ECR I-9693, at 42 and case law referred to therein.

A major part of case law on the practical application of Article 36 TFEU in the foodstuffs sector refers to pre-market authorizations of enriched foodstuffs.[44] *Commission* v. *Denmark* dealt with the issue of a Danish ban on foodstuffs to which vitamins and minerals were added. The Danish procedure of prior authorization was characterized by the existence of an administrative practice which made authorization possible only if the enrichment with vitamins or minerals was necessary to prevent a situation where a large part of the population has an insufficient intake of the nutrient in question. The Danish practice constituted an obstacle to the free movement of goods, since it banned products that were lawfully marketed in other Member States. Yet, the Danish authorities contended that the prohibition was necessary for the protection of public health, and therefore permitted under Article 36 TFEU.

EU law does not preclude Member States from maintaining prior authorization procedures for certain types of foodstuffs. However, Article 36 TFEU is an exception to the free movement of goods, and, as such, it has to be interpreted strictly. It is therefore for the national authorities to prove that their measures are necessary to assure the level of the protection of human life and health chosen in that Member State. For risk analysis, it means that a prohibition on the marketing on certain products must be based on a detailed, comprehensive risk assessment, carried out by the national authorities in response to an application for authorization. Risk assessment must precede the decision of the public authorities.[45] It must show that there is a real risk to public health, 'sufficiently established on the basis of the latest scientific data available at the date of the adoption of the decision'.[46] In the Danish case, the mere fact that there is no nutritional need of the population justifying the addition of nutrients to foodstuffs could not be the sole reason for the total prohibition of the marketing of fortified foodstuffs. The measure was disproportionate because it consistently prohibited all foodstuffs to which minerals or vitamins were added, without referring to more detailed studies about the level of risk that they might pose to human health. The identification and assessment of a real risk to public health requires a case-by-case assessment.

The Commission's Communication on the precautionary principle states that prior approvals reverse the burden of proof by requiring that some substances, such as food additives or pesticides, be deemed hazardous until proven otherwise (EC, 2000b). This obligation is particularly relevant to measures where the precautionary

---

[44] See Case 174/82, *Sandoz*, [1983] ECR 2445; Case 192/01, *supra* note 43; Case 95/01, *John Greeham and Léonard Abel*, [2004] ECR I-1333; Case C-24/00, *Commission* v. *France*, [2004] ECR I-1277; Case 41/02, *Commission* v. *Netherlands*, [2004] ECR I-11375, Case 270/02, *Commission* v. *Italy*, [2004] ECR I-1559. National rules on adding vitamins and minerals to foodstuffs have now been harmonised – see Reg. 1925/2006 on the addition of vitamins and minerals and certain other substances to foods, OJ 2006, L 404/26.
[45] See Case E-3/00 *EFTA Surveillance Authority* v *Norway,* EFTA Court Report 2000/01: 73, at 34-42.
[46] Case 192/01, *supra* note 43, at 48.

principle is evoked. If supplementary scientific information is necessary for a more comprehensive risk assessment, the responsibility for financing further research is shifted onto professionals who have economic interest in pursuing further studies in order to show that the product can be granted market authorization (EC 2000b: 22). The producer is responsible, however, only for producing scientific evidence and carrying out scientific work needed to evaluate the risk. In *Sandoz*, also relating to the pre-market authorizations of foodstuffs to which vitamins have been added, the European Court of Justice confirmed that, although the national authorities may ask the producer or importer of a product to provide information relating to the compositional and technical characteristics of the product,[47] they must themselves assess, in the light of all available information, whether the product is not harmful. EU law does not permit national measures imposing a requirement on the producer to prove that a product is safe. The strict interpretation of the exceptions to the free movement of goods in Article 36 TFEU places the burden of assessing risks on the national authorities, which must, in each case, check whether the national measure complies with the criteria laid down by EU law.[48]

2.4.2 National derogations from harmonised measures

If laws are harmonised at the Community level, the national legislator can maintain only national provisions existing before the harmonisation measure (Article 114(4) TFEU) or introduce provisional national measures under safeguard clauses foreseen in the EU measure.

For the derogation contained in Article 114(4) TFEU, a notification procedure is mandatory. The Commission, within six month after receipt of a notification, verifies whether the national measure is a means of arbitrary discrimination or a disguised restriction on trade between Member States, and whether or not it constitutes an obstacle to the functioning of the internal market. If the Commission does not take a decision within this period (which, if need be, can be further extended for a period up to six months, in the case of the complexity of the matter and provided that no danger to human health exists), the notified provisions shall be deemed to have been approved (Article 114(6) TFEU). It should be stressed that, unlike in the case of measures in the non-harmonised area, the notification procedure for national measures derogating from the harmonized legislation does not only have an informative character, but requires approval by the Commission – a Member State is not authorized to apply its provisions until a decision from the Commission confirms them.[49]

---

[47] Case 174/82, *Sandoz, supra* note 44, at 25.
[48] *Ibid.*, at 22.
[49] Case 319/97, *Kortas*, [1999] ECR I-3143.

Member States have to provide the Commission with supporting information concerning the grounds for maintaining or introducing the national measures. As opposed to Article 114(5) TFEU, which requires (*sine qua non*) that new national measures be based on new scientific findings and concern a problem specific to that Member State,[50] national measures that predate harmonization are not subject to these conditions. Under Article 114(4) TFEU a Member State can maintain its national provisions even if the divergent national and EU measures are based on the same scientific evidence, which is assessed differently.[51] Nevertheless, in the same judgment, the Court confirmed that new scientific findings and a problem specific to the Member State can still be highly relevant factors during the approval procedure for measures based on Article 114(4) TFEU.[52]

The national risk management measures derogating from EU harmonised measures must guarantee a higher level of protection of human life and health than the EU measures. The obligation to prove that the level of protection is higher and that the national measures do not go beyond what is necessary to attain this objective falls on the Member State.[53]

2.4.3 Safeguard clauses

Safeguard clauses are included in most harmonization measures in the field of food safety. They allow Member States to restrict provisionally or to suspend the trade in and use of the food in question in their territory. A Member State invoking the safeguard clause must have detailed grounds for considering that the product may be dangerous to human life or health. The Commission examines the grounds as soon as possible and takes appropriate measures. The national measures may remain in force until the EU measures are approved.

The existing safeguard clauses are being gradually replaced by reference to Articles 53 and 54 GFL, which introduce a single emergency procedure applicable to all food and feed. This safeguard clause confers special powers on the Commission to take – at the request of a Member State or on its own initiative – emergency measures where it is evident that a product is likely to constitute a serious risk to human, animal or plant health, and that such risk cannot be contained satisfactorily by measures taken by the Member States. If, after receiving information from a Member State on the need to take emergency measures, the Commission does not act, the Member State may adopt provisional protective measures, which can be maintained until an EU decision concerning the said measure is adopted.

---

[50] Joint Cases T-366/03 and T-235/04, *Land Oberösterreich* and *Austria* v. *Commission*, [2005] ECR II-4005.
[51] Case C-3/00, *Denmark* v. *Commission*, [2003] ECR I-2643, at 63.
[52] *Ibid.*, at 59-61.
[53] *Ibid.*, at 64.

2.4.4 Risk analysis and proportionality

Exceptions to the principle of the free movement of goods have to be 'justified', which means that they have to be necessary for the desired objective and proportional. The standard of review applied by EU Courts to judge the legality of food safety measures introduces the obligation for Member States to provide scientific evidence, showing that a measure is necessary to protect public health. On the basis of the scientific evaluation the EU Courts assess whether the exercise of discretion in introducing national measures does not violate the principle of proportionality.

Proportionality requires that if a choice between various risk management options aiming at the same objective is possible, a Member State has to choose the least trade hindering one. A litmus test for the proportionality of national measures in the field of food safety is often adequate information. If the consumer can be sufficiently protected by appropriate labelling, a product should be admitted on the market and any measures going beyond this trade restriction should be considered unlawful.[54]

The Danish case discussed above provides another example of a measure judged unproportional. The systematic prohibition of marketing of all foodstuffs to which vitamins and minerals were added was not underpinned by a detailed assessment (case-by-case), distinguishing different minerals and vitamins and the effect on human health which their addition may cause. Because no scientific evidence existed to prove that the ban was necessary for the protection of the interests referred to in Article 36 TFEU, the mere absence of a nutritional need in the Danish population did not justify a total import ban on enriched foodstuffs.[55]

2.5 Risk analysis at EU level

2.5.1 Treaty provisions

The General Food Law provides a basic framework of overarching principles for food legislation at both EU and national levels. Therefore, the requirement that food law shall be science-based has also implications for the EU legislator.

The most important Treaty provisions for the field of food safety are contained in Article 114 TFEU. The Article explicitly states that approximation of laws covers areas concerning 'health, safety, environmental protection and consumer protection'. Article 114(3) TFEU requires that *a high level of protection* be ensured

---

[54] See, e.g. Case 178/84, *Commission v. Germany,* [1987] ECR 1227; Case 347/89, *Freistaat Bayern* v. *Eurim-Pharm GmbH,* [1991] ECR I-1747.
[55] Case 192/01, *supra* note 43, at 46 and 56; see also Case E-3/00, *supra* note 45, at 73.

in harmonisation legislation. The high level of protection does not necessarily have to be the highest technically possible. The final risk management decision depends on cost-benefit analysis, although – according to the established case law of the European Courts – the protection of human life and health takes precedence over economic considerations.[56]

Clearly, the European Union strives to ensure a high level of health protection. Some Member States expressed fears that giving up the veto power in the area of approximation of laws for the purpose of creating the internal market in consequence would force them to lower their national standards referring to the environment or health protection. Hence, the Treaty contains several mechanisms to ensure a level of health protection satisfactory for all Member States.

The Commission in its proposals concerning health, safety, environmental protection and consumer protection must take account of *any new development based on scientific facts*. This requirement was introduced by the Treaty of Amsterdam, and, as well as other amendments concerning the area of public health, was largely motivated by the BSE affair (Vos, 2000a). Moreover, Article 114(7) TFEU imposes on the Commission a duty to examine whether the relevant EU provisions need revision *immediately* after approving national provisions derogating from it. A similar requirement of prompt reaction from the Commission is contained in Article 114(8) TFEU, where the Commission has to examine *immediately* whether to propose adequate measures at EU level, if a Member State raises a specific problem on public health in a field which has been subject of harmonisation measure. The latter is a consequence of the lack of the possibility to introduce measures on grounds of the exceptions listed in Article 36 TFEU by the Member States themselves. The rationale behind this provision is to assure that the EU legislature will respond quickly and effectively to problems occurring in different Member States by adapting the harmonised legislation to new circumstances.

2.5.2 EU scientific advice

Drawing up new legislation in the field of food safety and execution of the existing legislation require recourse to complex and technical scientific risk assessments. To ensure a high level of health protection, measures must be based on the best available science and risk assessment must include different scientific views and scientific uncertainty. The principle of science-based food law applied at EU level raises two important issues: which scientific opinion the EU institutions should follow, and to what extent they can deviate from risk assessment conclusions.

---

[56] Case T-304/01, *Julia Abad Pérez and Others* v. *Council and Commission*, [2006] ECR II-4857, at 60 and case law referred to therein.

The issue of risk analysis, the precautionary principle and the use of scientific advice by EU institutions was dealt with in the *Pfizer* and *Alpharma* judgments. The judgments concerned the withdrawal of authorizations for two antibiotics, virginiamycin and bacitracin zinc, used in animal feedingstuffs as growth promoters.[57] The EU justified this decision on grounds of the protection of public health. Both antibiotics were also used in treating human infections, and their use in feedingstuffs could result in developing resistance to these antibiotics in humans. Because this possibility could not be established with scientific certainty, the Council evoked the precautionary principle to justify its measure.

Before the European Food Safety Authority (EFSA) was established, risk assessment in the field of food safety had been carried out by five scientific committees[58] and the Steering Committee. In May 2003 EFSA took over their tasks. EFSA is an advisory body responsible for delivering scientific advice to EU institutions. In the risk analysis methodology, scientific bodies provide a reasoned analysis of the case in the light of all relevant knowledge about the subject, which allows public authorities to take an informed decision. However, risk assessors have neither political responsibilities nor democratic legitimacy to exercise public authority. Therefore, EFSA is not responsible for choosing a level of protection that is acceptable for society and for selecting risk management measures that ensure that level.

In practice, it means that public authorities are not bound to follow risk assessment conclusions.[59] In *Pfizer*, EFSA's predecessor – the Scientific Committee on Animal Nutrition (SCAN) – was asked to provide an opinion on the use of virginiamycin, taking into account the scientific evidence, produced by the Danish authorities, concerning the transfer of antimicrobial resistance from animals to humans. SCAN concluded that 'the use of virginiamycin did not constitute a real immediate risk to public health'. However, the EU institutions disregarded this conclusion in their decision, even if the decision was based on various items of the scientific evidence contained in the SCAN's opinion.[60]

Interestingly, in this context the Court set several conditions for the EU institutions opting to disregard scientific conclusions provided by advisory bodies. Firstly, they must provide specific reasons why the scientific opinion provided by the advisory body is not followed. The explanation of reasons must be of a scientific level at least equal to the rejected opinion. It can be either a supplementary

---

[57] Case T-13/99, *Pfizer Animal Health SA v. Council*, [2002] ECR II-3305; Case T-70/99, *Alpharma v. Council*, [2002] ECR II-3495.
[58] These committees were: Scientific Committee on Food, Scientific Committee on Animal Nutrition, Scientific Veterinary Committee, Scientific Committee on Pesticides, and Scientific Committee on Plants.
[59] *Pfizer, supra* note 57, at 196.
[60] *Ibid.*, at 200.

opinion from the same committee or other evidence, whose 'probative value is at least commensurate with that of the opinion concerned'. If the EU institution decides to follow the opinion only partly, it may refer in its statement of reasons the parts of the opinion which it does not dispute (in *Pfizer*, the Commission justified its measure by reference to three recitals of SCAN's opinion, which in its view constituted a sufficient proof of the existence of risk).[61]

The judicial review of measures based on risk assessment is in general confined to examining whether the exercise of discretion by the European institutions is vitiated by a 'manifest error or a misuse of powers' and whether the institutions clearly exceeded the bounds of their discretion'.[62] Türk suggests that the EU Courts distinguish between basic acts where the EU legislature enjoys a broad discretion, and implementing acts where the administrative discretion is the subject of a more detailed judicial review (Türk, 2006). However, as shown in *Pfizer*, the judicial review of implementing acts of general application is also limited.

The Community Courts are generally reluctant to re-evaluate the scientific assessment of highly complex scientific and technical facts or choices of political and social nature where different interests have to reconciled. The policy choices are the political responsibility of the EU institutions, on which the Treaty confers this duty, and therefore the Courts do not engage themselves in substituting their own assessment for the assessment on which the EU institutions relied.[63] As the Court stated in *Artegodan* – a case concerning medicinal products – only the proper functioning of the relevant scientific committee, the internal consistency of the opinion and the statement of reasons contained therein are subject to judicial review.[64] Furthermore, regarding the statement of reasons, the Court is empowered only to examine whether it is possible to ascertain from it 'the considerations on which the opinion is based', and whether a comprehensive link between the scientific findings and its conclusions exists. In the case of a significant discrepancy, the scientific body has to state the reasons why it has departed from the conclusions or expert opinions, in order to ensure a detailed and objective risk assessment, based on the most representative opinions. This requirement is particularly important in the case of scientific uncertainty.

According to the General Food Law, food safety measures should be based on risk analysis. Does it mean that recourse to scientific advice is necessary before any food safety measure is taken? In the Angelopharm judgment, concerning cosmetic products, the ECJ found that the EU institutions did not have the necessary expertise to assess scientific findings on their own and therefore – in order to ensure

---

[61] *Pfizer, supra* note 57, at 199.

[62] *Ibid.*, at 166.

[63] *Ibid.*, at 169; Case T-19/01, *Chiquita*, [2005] ECR II-315, at 228.

[64] Joined Cases T-74/00, T-76/00, T-83/00, T-84/00, T-85/00, T-132/00, T-137/00 and T-141/00, *Artegodan and Others* v. *Commission*, [2002] ECR II-4945. at 200.

that 'measures adopted at EU level are necessary and adapted to the objective of protecting human health' – consultation of the relevant Scientific Committee must be mandatory in all cases.[65] In a broad sense, this would require the EU institutions to consult Scientific Committees (in the area of food law – EFSA) every time they take a decision requiring scientific or technical knowledge.

Pfizer cast some light on the obligation to consult scientific committees established in *Angelopharm*. The judgment narrowed down the Court's findings in *Angelopharm* to the interpretation of the Directive in question, and did not see them applicable in the Pfizer case. However, the Court highlighted that the scientific evidence available to the EU institutions must be 'sufficiently reliable and cogent for them to conclude that there was a risk'.[66]

In the area of food and feed, EFSA is designed to be a scientific point of reference for the EU institutions, ensuring a high scientific quality and enjoying the confidence of general public and other interested parties. However, consulting EFSA is obligatory only if legislation explicitly provides for that. The Regulation establishing a common authorisation procedure for food additives, food enzymes and food flavourings[67] is an example of a centralized risk assessment carried out under the responsibility of EFSA. The Commission is required to seek EFSA's opinion unless the procedure concerns an issue not related to human health. Furthermore, the Regulation recognizes that risk management may include other legitimate factors, and therefore the Commission may propose a measure that is not in line with EFSA's opinion. However, in this case the Regulation imposes an obligation to explain the reasons for the decision,[68] which is in line with EU case law.

2.6 Precautionary principle

2.6.1 Precautionary risk management and risk assessment

The General Food Law contains in Article 7 a definition of the precautionary principle. This is the first time the principle has been defined in EU law. Previously, the application of the precautionary principle to different areas of EU activity

---

[65] Case 212/91, *Angelopharm* v. *Freie Hansestadt Hamburg*, [1994] ECR I-171, at 38.
[66] *Pfizer, supra* note 57, at 322.
[67] *Supra* note 27.
[68] Art. 7(3) Reg. 1331/2008.

was based on the Treaty, which only adopts the principle, without determining conditions of putting it into practice.[69]

The precautionary principle is part of the risk analysis methodology (Belvèze, 2003). It allows decision makers to act without having to wait until the reality and seriousness of risk to health are fully demonstrated. The principle is a provisional risk management tool, 'pending further scientific information for a more comprehensive risk assessment'.[70] Hence, the principle is closely connected to risk assessment and should be preceded by a comprehensive evaluation of possible risks to human health based on the most recent scientific information.[71]

Measures may be based on the precautionary principle if the risk, 'although the reality and extent thereof have not been 'fully' demonstrated by conclusive scientific evidence, appears nevertheless to be adequately backed up by the scientific data available at the time when the measure was taken'.[72] Precautionary measures cannot be based on 'purely hypothetical or academic considerations', founded on 'mere suppositions which are not yet scientifically verified'.[73] Yet, this requirement is ample enough to encompass precautionary measures introduced in the case of diverging scientific opinions or measures based on some parts of risk assessment that, in the view of the risk managers, constitute sufficient information to conclude that a possible risk to human health exists. As shown in *Pfizer*, the precautionary measures can be legitimately introduced despite scientific conclusions stating no risk to human health.[74]

At the same time, however, the European Courts impose some limitations on the considerable discretion that public authorities enjoy in the situation of scientific uncertainty. The *Artegodan* judgment, already referred to in the previous section, dealt with the withdrawal of authorization of several products containing amphetamine-like anorectics accelerating the feeling of satiety and used in the treatment of obesity. The reason for withdrawal, according to the Commission, concerned changes in good clinical practices, requiring the assessment of long-

---

[69] Art. 191(2) TFEU states: 'Union policy on the environment shall aim at a high level of protection. It shall be based on the precautionary principle'. In *Artegodan*, the CFI confirmed that – since the requirement of the protection of public health, safety and environment applies to all spheres of EU activity, the precautionary principle has a status of an autonomous principle in EU law (*supra* note 64, at 183-184). The interpretation derives from Art. 191(1) TFEU ('Union policy on the environment shall contribute to pursue the objective of protecting human health') and from Art. 11 TFEU, which stipulates that environmental protection requirements must be integrated into the definition and implementation of the Union policies and activities.
[70] Art. 7(1) GFL.
[71] Case E-3/00, *supra* note 45, at 30.
[72] *Pfizer*, *supra* note 57, at 144.
[73] See Case E-3/00, *supra* note 45, at 29; Case 236/01, *Monsanto Agricoltura Italia and Others*, [2003] ECR I-8105, at 106.
[74] For comments on the precautionary principle in the *Pfizer* judgment see Da Cruz Vilaça (2004) and De Sadeleer (2006).

term effects of medicinal products in the treatment of obesity. The previous assessment confirmed evidence of short-term therapeutic effects, but did not prove the long term efficacy of the products. The Court of First Instance (now renamed the General Court) held that mere changes in the perception of risk (in this case changes in good clinical practices) cannot justify a precautionary decision to withdraw authorization for medicinal products. The withdrawal decision was annulled because it was not based on any new scientific data or information.[75] Thus, the withdrawal of a marketing authorization can in principle be justified only where a potential risk or the lack of efficacy is proven by new scientific evidence, unless the exceptional situation occurs in which public authorities acknowledge that they had incorrectly assessed a product.[76] This also applies to the change or introduction of new assessment criteria – they can affect the authorisation's validity only if the development is based on new scientific information.

Within the internal market, Member States can apply the precautionary principle to measures based on Article 36 TFEU (exception to the principle of the free movement of goods on grounds of the protection of public health), and to national derogations from harmonization measures based on Article 114(4) TFEU and safeguard clauses existing in specific EU legislation.

For the European Union, the potential weakening of the internal market by recognizing the wide discretion of Member States in drawing up their food safety measures under scientific uncertainty is compensated by a wide regulatory discretion at European level. The way the principle is applied within the EU affects the position the EU defends internationally within the framework of the WTO SPS Agreement (EC, 2000b: 2), where it strives to bargain maximum discretion. Besides, regarding national precautionary measures, the Court has consistently held that, to the extent that uncertainties continue to exist in the current state of scientific research, it is for Member States to determine the level of protection of public health they deem appropriate.[77] Thus, the formal inclusion of the precautionary principle among the general principles of the EU food safety law does not change so drastically the discretionary power of Member States in cases of scientific uncertainty.

2.7 Conclusions

The requirement to base food safety measures on scientific evidence had been a standard of review applied by the EU Courts long before the principle of risk

---

[75] *Artegodan, supra* note 64, at 211.

[76] *Artegodan, supra* note 64, at 194.

[77] See, e.g. Case 174/82, *supra* note 44, at 16; Case 42/90, *Bellon*, [1990] ECR I-4863; Case C-192/01, *supra* note 43.

analysis was introduced in the General Food Law as one of the general principles of the new food safety policy.

The principle of risk analysis and the precautionary principle together constitute a methodology to be applied by the legislator at national and EU levels in drawing up measures in the field of food safety. Because the aim of this methodology is to ensure consistency in decision making, risk analysis imposes the same methods for both national and EU legislation.

On the basis of scientific risk evaluations, the EU Courts assess whether national measures based on Article 36 TFEU (exceptions to the free movement of goods) are proportional. Thus, from the internal market perspective, the principle of risk analysis applied to the non EU-harmonised area could be considered as a negative integration instrument,[78] since the requirement of scientific risk assessment for food safety measures sets limits for national legislators.

On the other hand, risk analysis and scientific governance create some elements of positive integration: an institution designed to be the scientific point of reference in food safety issues for the entire EU, and new instruments: the mediation procedure set out in Article 60 GFL and EFSA's obligation to cooperate with national bodies in case of diverging scientific opinions.

Without discussing here the myriad of political aspects of its practical application to EU multi-level food safety governance, this legal structure built upon the principle of risk analysis adds a new dimension to the reinforcement of the internal market. Because the judicial review of scientific evidence is limited to formal requirements, the close cooperation and integration of scientific expertise within the European Union have a theoretical potential to develop a common scientific position on food safety issues and thus to contribute to eliminating barriers to intra-EU trade. While mutual recognition is the EU answer to obstacles to the free movement of goods created by different national technical and quality standards, the approximation of scientific risk assessments offers a solution to different understandings of food safety.

---

[78] For the concept of negative and positive integration see Tinbergen (1965). The author defines *negative* integration as 'measures consisting of the abolition of a number of impediments to the proper operation of an integrated market', whereas *positive* integration means 'the creation of new institutions and their instruments or the modification of existing instruments'.

3. Science-based legislation?

EU food law submitted to risk analysis

Abstract

Food safety is the only area of risk regulation where a comprehensive risk analysis model has been introduced not only by international trade agreements, but also by EU legislation, as one of the general principles governing policy. This chapter analyses the scope of application of risk analysis and the precautionary principle in EU food safety regulation on the example of prior authorisation schemes. These schemes set up a legislative 'framework', under which regulatory decisions concerning the placing on the market of foodstuffs belonging to certain categories are being issued. Authorisation decisions are based on risk assessments. The question remains, however, whether the legislative framework itself should also be submitted to risk analysis. To what extent does the risk analysis methodology set limitations on the legislator in deciding on food safety measures that restrict trade? Are EU food safety measures legal under international trade obligations? Do they comply with the general principles of EU food law?

3.1 Introduction

Risk regulation, by its very nature, interferes with market processes to protect fundamental welfare of citizens, such as health or safety. The challenge national and regional (EU) politics face to balance free trade and health protection in this highly technological area consists in balancing choices made in an ideally deliberative environment of experts and technocrats with popular will and laypeople's risk perception. To be able to include in risk management a broad array of factors other than hard-core scientific facts, decision makers must be entrusted with a certain amount of discretion. From a trade perspective, however, such discretion may be used for protectionist purposes. Therefore, the system of trade rules aims to limit discretion as much as possible, by making 'scientific risk assessment' the criterion deciding which measures are justified.

One of the most visible areas of risk regulation where conflicts between popular choices and scientific evidence arise is food safety. The World Trade Organization (WTO) Agreement on Sanitary and Phytosanitary Measures (SPS) introduces the requirement that national SPS measures be based on risk assessment and not maintained without sufficient scientific evidence. Risk analysis has also become a general principle of EU food law, introduced by Regulation 178/2002 laying down the general principles and requirements of food law, establishing the European

Food Safety Authority and laying down procedures in matters of food safety[1] (the so-called General Food Law – GFL).

Obviously, the institutional context of risk regulation is predominantly public administration (Fisher, 2007). The growing demand for market regulation in modern welfare states and a corresponding need for the adoption of detailed rules referring to technical or scientific knowledge have led to the rise of government as law-maker. The vast majority of legal acts in the EU and at national level, ranging from the adoption of standards and market authorisations to the amendment of basic acts, are delegated acts (Türk, 2006).

Not all risk regulation, however, is delegated by primary law-maker to administration. The question that remains unanswered in this context is whether risk analysis applies only to authorization decisions and other administrative activities 'under legislation', or whether it should also bind the legislature. The EU internal market straitjacket clearly imposes limitations on legislators in the Member States, by requiring scientific proof for national food safety measures restrictive of intra-EU trade. Does risk analysis, however, set limitations on the EU legislator as well? Are EU food safety measures legal under international trade obligations?

This chapter analyses the scope of application of risk analysis and the precautionary principle in EU food safety regulation on the example of prior authorisation schemes. Prior authorisation schemes are legislative measures which consider certain categories of products a priori dangerous – until the manufacturer or exporter interested in marketing these products proves otherwise. To gain approval, the proponent must provide adequate scientific evidence to ensure that a product (or technology) is safe. Thus, prior authorisation schemes set up a *legislative* 'framework', under which *regulatory* (administrative) decisions concerning the placing on the market of foodstuffs belonging to certain categories are being issued. These individual decisions are based on risk assessment. Do the prior authorisation schemes themselves, however, require scientific justification? In other words, what is the scope of application of risk analysis and its impact on the discretion of EU legislator?

As a precursor to this discussion, we will first explain the structure of the EU food law. Then, the principle of risk analysis and the precautionary principle in both EU food law and in international trade regime will be presented in outline, with a focus on the precautionary principle as prior authorisation schemes are often referred to as precaution-inspired. Next, we will put the EU prior authorisation schemes to two tests. Firstly, we will refer to the recent case-law of the European Court of Justice on the legality of a French prior authorisation scheme where the principle of risk analysis was referred to. We will analyse whether the Court's

[1] OJ 2002, L 31/1.

interpretation can also be applied to legislation at EU level. Secondly, the EU food safety legislation will be examined in the context of the obligations imposed by the WTO SPS Agreement.

3.2 EU food safety regulation

3.2.1 EU General Food Law

The entry into force of Regulation No 178/2002 is often characterised as the creation of a new comprehensive EU policy (Lafond, 2001; Szajkowska, 2009; Ugland and Veggeland, 2006). The Regulation established, i.e. an EU definition of food, an overarching requirement that food shall not be placed on the market if it is unsafe, a traceability system, a Rapid Alert System for Food and Feed (RASFF), emergency procedures, and defined public and private responsibilities for assuring food safety. Articles 5-8 GFL set out the following general principles of food law:

- Food law shall pursue the general objectives of a high level of protection of human life and health and the protection of consumers' interests, including fair practices in food trade, taking account of, where appropriate, the protection of animal health and welfare, plant health and the environment; while aiming to achieve the free movement of food and feed in the EU (Article 5).
- Food law shall be based on risk analysis (Article 6).
- In specific circumstances, the precautionary principle may be adopted (Article 7).
- Food law shall aim at the protection of consumers' interests and shall provide a basis to make informed food choices by aiming at the prevention of fraudulent and deceptive practices, the adulteration of food, and any other practices which may mislead the consumer (Article 8).

These general principles apply to all stages of production, processing and distribution of food and feed for food-producing animals, leaving outside their scope only primary production for private domestic use and the domestic preparation, handling or storage of food for private domestic consumption.[2] Given that 'food law' is defined as 'laws, regulations and administrative provisions governing food in general, and food safety in particular, whether at Union or national level',[3] the Regulation forms a truly general framework covering all fundamental aspects of the food safety policy, of a horizontal nature applied to all measures concerning food or feed, both at EU and national levels.

Figure 3.1 shows the structure of EU food law. Regulation 178/2002 applies to all measures taken in the EU multi-level food safety governance. Horizontal provisions of food law exist under the umbrella of the General Food Law. These provisions refer to legislation that covers as broad a category of foodstuffs as possible. An

---

[2] Art. 1(3) GFL.
[3] Art. 3(1) GFL.

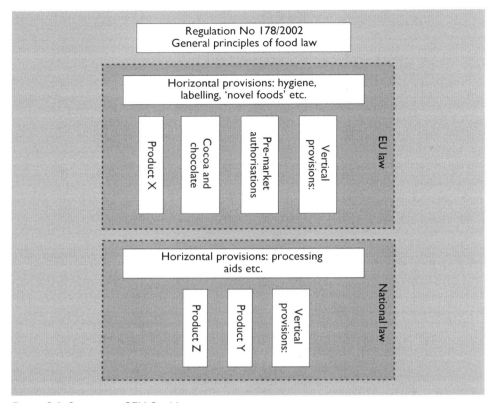

Figure 3.1. Structure of EU food law.

example of horizontal legislation is the Labelling Directive – its rules of a general nature are applicable to all foodstuffs put on the market.[4] Rules of a specific nature which apply only to particular foodstuffs should be laid down in provisions dealing with those products.[5]

Vertical provisions are specific measures concerning certain foodstuffs. These are a mix of legislative and regulatory measures delegated to public administration. An example of a legislative act is the Directive relating to cocoa and chocolate products,[6] setting the cocoa percentage for chocolate, milk chocolate, white chocolate, pralines, etc. in the whole European Union. Chocolate marketed in the EU territory not only has to comply with the Labelling Directive and other horizontal provisions, but – additionally – with specific 'recipe' provisions relating to its compositional standard.

---

[4] OJ 2000, L 109/29.
[5] Rec. 4-5 Dir. 2000/13.
[6] OJ 2000, L 197/19.

Regulation 1829/2003 on genetically modified food and feed[7] is an example of a *horizontal* act applying to all products containing, consisting of or produced from genetically modified organisms (GMOs). GMOs must undergo a safety assessment through an EU procedure before being placed on the EU market. Hence, within this 'framework', regulatory decisions concerning individual GM products, such as GM maize[8] or soybean[9] are being issued (vertical decisions 'under legislation'). A similar prior authorisation procedure exists for novel foods. The procedure will be discussed in more detail later in this chapter.

Finally, the general principles and requirements established in the General Food Law also apply to the food laws of the 27 Member States. At national level, the same distinction between horizontal and vertical provisions applies. These national legal systems, however, have to be filtered through the Treaty provisions relating to free movement of goods. In light of Articles 34-36 of the Treaty on the Functioning of the European Union (TFEU), national measures prohibiting or restricting imports, exports or goods in transit on grounds of the protection of health and life of humans, animals or plants are exceptions to the rule of free movement of goods within the EU, which is one of the fundamental principles of the internal market. Because national measures aimed at the protection of human life and health are allowed by the Treaty, food safety provisions are to a large extent harmonised at EU level to avoid trade barriers. Through harmonisation (Article 114 TFEU), the EU legislator is gradually taking over Member States' competences in the field of food safety.[10]

Apart from national measures aimed at health, consumer or environmental protection, a number of compositional or technical standards relating to foodstuffs exist in national laws. As these standards fall outside the scope of the exceptions allowed by the Treaty, they cannot constitute a barrier to intra-EU trade. They are not aimed at the protection of human life or health, hence the principle of risk analysis does not apply to them either. The mutual recognition principle guarantees free movement of foodstuffs in these areas. According to the mutual recognition mechanism, goods which are lawfully produced in one Member State cannot be banned from sale in the territory of another Member State, even if the technical or quality specifications of these goods do not comply with the national

---

[7] OJ 2003, L 268/1.

[8] Dec. 2007/701 authorising the placing on the on the market of products containing, consisting of, or produced from genetically modified maize NK603xMON810 (MON-ØØ6Ø3-6xMON-ØØ81Ø-6), OJ 2007, L 285/37.

[9] Dec. 2008/730 authorising the placing on the market of products containing, consisting of, or produced from genetically modified soybean A2704-12 (ACS-GMØØ5-3), OJ 2008, L 247/50.

[10] In relation to Spain, Palau estimates that, in some years after the completion of the internal market, when positive integration in the field of food safety increased significantly, the percentage of food safety legislation totally or partially defined by EU acts reached 90% (Palau, 2009).

standards. Thus, mutual recognition ensures free movement of goods within the EU without the need to harmonise all national compositional standards.

3.2.2 International food law: a meta-framework

The technocratic character of decision-making processes grows as policy areas become subject to globalisation processes (UNRISD, 2004). In the global arena different players contribute to international food law. The Codex Alimentarius Commission, established by the Food and Agriculture Organization (FAO) and the World Health Organization (WHO), is a body developing food standards, guidelines and other texts such as codes of practice with main aims to protect the health of consumers and to ensure fair practices in international food trade (FAO/WHO, 2002, 2006; Knudsen *et al.*, 2008; Masson-Matthee, 2007).

Codex standards are voluntary and not legally binding on member countries. To acquire legal effect, these standards need to be implemented into national or regional (EU) legal systems.[11] Nevertheless, the Codex Alimentarius has relevance to the international food trade. The universally agreed upon and uniform food standards serve as a reference point for international disputes concerning food safety and consumer protection and as a benchmark against which national food safety measures are judged.[12]

Particularly the WTO Agreements on the Application of Sanitary and Phytosanitary Measures (SPS Agreement) and on Technical Barriers to Trade (TBT Agreement) encourage the adoption of Codex standards by national legislators. SPS measures that conform to international standards are deemed to be necessary to protect human, animal or plant life or health. In other words, measures that conform to standards promulgated by the Codex Alimentarius Commission enjoy a presumption of compliance with the WTO regime and are by definition considered necessary.

The SPS Agreement thus creates a strong incentive to follow Codex standards. Member states may, however, depart from international standards and introduce or maintain SPS measures which result in a higher level of protection than would

---

[11] Art. 13 GFL stipulates that the EU and its Member States shall '(a) contribute to the development of international technical standards for food and feed and sanitary and phytosanitary standards; (b) promote the coordination of work on food and feed standards undertaken by international governmental and non-governmental organisations; (c) contribute, where relevant and appropriate, to the development of agreements on recognition of the equivalence of specific food and feed-related measures; (d) give particular attention to the special development, financial and trade needs of developing countries, with a view to ensuring that international standards do not create unnecessary obstacles to exports from developing countries; (e) promote consistency between international technical standards and food law while ensuring that the high level of protection adopted in the Community is not reduced'.

[12] At present, Codex comprises more than 200 standards, close to 50 food hygiene and technological codes of practice, some 60 guidelines, over 1,000 food additives and contaminants evaluations and over 3,200 maximum residue limits for pesticides and veterinary drugs.

be achieved by measures based on international standards. They must provide scientific justification for the stricter measures. Scientific evidence is not required only when food safety measures are based on Codex standards.

The SPS Agreement introduces a science-based regime governing international trade in agricultural products and foodstuffs. Article 2(2) of the Agreement states that WTO members shall ensure that any SPS measure is 'based on scientific principles and not maintained without sufficient scientific evidence'. Article 5 SPS introduces risk assessment as a method for determining the appropriate level of health protection, carried out according to the internationally developed techniques.

All standards concerning food safety that serve as a benchmark are developed outside the WTO. The SPS Agreement is confined to a purely procedural requirement that sanitary or phytosanitary measures be based on scientific evidence. Hence, the SPS Agreement sets out a framework of requirements that apply to national legal systems and therefore it can be considered a meta-framework (Van der Meulen, 2010a).

Because WTO agreements bind member states, the WTO plays a role in the application of food standards in trade and in dispute resolution. The Dispute Settlement Understanding provides a procedure to resolve conflicts. If a party so requires, the Dispute Settlement Body (DSB) forms a panel to deal with the issue. Panel decisions can be appealed to the Appellate Body (AB). The WTO cannot enforce decisions taken in this procedure. It can, however, allow the winning party to implement economic sanctions if the party found at fault does not comply. These sanctions are usually additional import levies on goods from the member state found at fault.

3.2.3 The risk analysis principle

The requirement that food safety measures be science-based and the risk analysis methodology are also set out in Regulation 178/2002 (Table 3.1). The precautionary principle, which allows taking provisional food safety measures where a risk to life or health exists but scientific uncertainty persists, is codified in the same Regulation and linked to the framework of risk analysis (Szajkowska, 2010).

Table 3.1. Risk analysis methodology applied to EU and national measures.

	SPS requirements	**EU requirements**
EU food law	based on risk assessment	based on risk assessment and other legitimate factors
	based on Art. 5.7 SPS in case of uncertainty (temporary measures)[13]	based on the precautionary principle in case of scientific uncertainty (temporary measures)
National food law	based on risk assessment	based on risk assessment and other legitimate factors
	based on Art. 5.7 SPS in case of uncertainty (temporary measures)	bBased on the precautionary principle in case of scientific uncertainty (temporary measures)

Similarly to the widely accepted definition elaborated by the Codex Alimentarius Commission, risk analysis established by the General Food Law consists of three components: risk assessment, risk management and risk communication. Science-based regulatory actions require a scientific risk assessment. Its aim is to identify and characterize a hazard, to assess exposure and to characterise risk. Risk assessment must be based on the available scientific evidence, and carried out in an independent, objective and transparent manner.[14]

Based on the outcomes of risk assessment, a process of weighing policy alternatives leads to – if need be – the selection of appropriate prevention and control options, which are designed to keep risk at a level acceptable to society. The second step – risk management – takes into account risk assessment, but also other legitimate factors, such as societal, economic, traditional, ethical or environmental concerns.[15]

Finally, the aim of the third element of risk analysis – risk communication – is the interactive exchange of information and opinions throughout the risk analysis process as regards hazards and risks, risk related factors and risk perceptions, among risk assessors, risk managers, consumers, businesses, and all other interested parties.[16]

---

[13] Art. 5.7 SPS reads: 'In cases where relevant scientific evidence is insufficient, a Member may provisionally adopt sanitary or phytosanitary measures on the basis of available pertinent information, including that from the relevant international organizations as well as from sanitary or phytosanitary measures applied by other Members. In such circumstances, Members shall seek to obtain the additional information necessary for a more objective assessment of risk and review the sanitary or phytosanitary measure accordingly within a reasonable period of time.'
[14] Art. 6(2) GFL.
[15] Rec. 19 GFL.
[16] Art. 3(13) GFL.

It has to be noted that – contrary to Codex Alimentarius and the EU General Food Law – the SPS Agreement mentions only risk assessment and remains silent about the other components of risk analysis and risk analysis itself. It is recognized, however, that risk assessment referred to in the SPS Agreement corresponds to the risk analysis methodology and its three steps (Belvèze, 2003). Risk management, although not defined per se in the Agreement, is reflected in the concept of 'sanitary and phytosanitary measures', which are the result of decisions on actions to reduce or eliminate risks presenting more danger than society is willing to accept.[17]

3.2.4 The precautionary principle

The precautionary principle is considered a key tenet of the EU food safety governance. The General Food Law contains a definition of the principle applicable to all food legislation at EU and national levels. In accordance with Article 7 GFL, the precautionary principle allows taking provisional measures where the possibility of harmful effects on health has been identified but scientific uncertainty persists. The General Food Law sets out the following conditions to be satisfied if the precautionary principle is to be considered:
1. available scientific data has to be assessed before a decision is made;
2. potentially dangerous effects deriving from a phenomenon, product or process, have to be identified;
3. the scientific evidence is inconsistent or inconclusive and does not allow the risk to be determined with sufficient certainty.

Measures based on the precautionary principle are provisional, pending more scientific information, and have to be reviewed within a reasonable period of time. They also have to be proportional and no more restrictive of trade than is required to achieve the high level of health protection.

The SPS Agreement contains one, indirect reference to a precautionary approach, as an exception to the requirement that SPS measures are based on risk assessment and not maintained without sufficient scientific evidence. The exception is contained in Article 5.7 SPS Agreement and applies to cases where:
1. relevant scientific evidence is insufficient;
2. measures are adopted on the basis of available pertinent information.

---

[17] In *EC-Hormones*, the Panel made an effort to distinguish between risk assessment as a 'scientific' examination of data and factual studies, and risk management defined as a 'policy' exercise involving social values and judgements. The Appellate Body, however, refrained from discussing the findings of the Panel as it found no textual basis for this distinction – see WTO Appellate Body Report, EC – Measures Concerning Meat and Meat Products, WT/DS26/AB/R; WT/DS48/AB/R, adopted 13 Feb 1998 (*EC-Hormones*), para. 181.

Such measures have to be reviewed within a reasonable period of time and member states must seek to obtain additional information necessary for a more objective assessment of risk. All these elements are cumulative: whenever one of them is not met, the measure is inconsistent with Article 5.7 SPS.[18]

Although the elements of the precautionary principle in the General Food Law and precautionary measures allowed under Article 5.7 SPS Agreement look very similar (if not the same), the EU policy makers seem to bestow upon the principle a broader scope of application.

3.2.5 Stretching the precautionary principle

The discussion on the precautionary principle has often been raised within the WTO and the Codex Alimentarius Commission, causing controversy between the EU – strongly advocating the application of the precautionary principle in food safety policy – and other countries, fearing that the EU would use the principle to justify regulatory decisions based on factors other than scientific evidence (CAC Allio *et al.*, 1999, 2006; Wiener and Rogers, 2002).

The controversy largely comes down to the place and role of scientific uncertainty in the risk analysis process, and – more precisely – whether the principle is linked to risk assessment or risk management. In this regard, the European Commission in its Communication on the Precautionary Principle distinguishes between a *prudential approach* applied by scientists and the application of the *precautionary principle* pertaining to the realm of politics and explains the distinction as follows:

> These two aspects are complementary and should not be confounded. The prudential approach is part of risk assessment policy which is determined before any risk assessment takes place ... it is therefore an integral part of the scientific opinion delivered by the risk evaluators. On the other hand, application of the precautionary principle is part of risk management, when scientific uncertainty precludes a full assessment of the risk and when decision makers consider that the chosen level of environmental protection or of human, animal or plant health may be in jeopardy (EC, 2000b: 13).

The General Food Law follows this approach. Article 6(3) explicitly refers to the precautionary principle as one of the elements that should be taken into account in risk management, apart from risk assessment and other factors legitimate to the matter under consideration. The wording of this Article confirms that the precautionary principle in the EU is clearly considered as a risk management

---

[18] WTO Appellate Body Report, Japan – Measures Affecting Agricultural Products, WT/DS76/AB/R, adopted 19 Mar 1999 (*Japan – Agricultural Products II*), para. 89.

tool, and not as uncertainties built in the risk assessment and dealt with in the scientific conclusions.

Although the *Working Principles for Risk Analysis for Food Safety for Application for Governments* – a guidance to national governments for food safety measures, negotiated within the Codex Alimentarius Commission and adopted in 2007 – entrusts the responsibility for resolving the impact of uncertainty on the risk management decision on the risk managers, and not on the risk assessors (CAC, 2007b: para. 28), Codex also states that the assumptions used for the risk assessment and the risk management measures selected should reflect the degree of uncertainty. In the same way, it highlights that the decisions should be based on risk assessment, and should be proportionate to its results, taking into account, where appropriate, other factors legitimate to the matter under consideration.[19]

Thus, although the Codex guidelines consider precaution an inherent element of risk analysis and even confirm that uncertainty exists during both risk assessment and risk management (CAC, 2007b: para. 12), they concentrate on linking the principle as much as possible with scientific evidence, reiterating the importance to present explicitly, in a transparent manner, all constrains, uncertainties and assumptions that have an impact on risk assessment.

Conversely, the efforts of the EU legislator seem to be directed at detaching the precautionary principle from risk assessment. As a risk management tool, the principle allows decision makers to overrule the findings of risk assessment, and thus broadens their discretion. Under the head of the precautionary principle, the EU legislator can still take measures contrary to the overall conclusion of the technical and scientific assessment showing that risks are acceptable.[20] We will now turn to the example of prior authorisation schemes as one of the EU applications of the precautionary principle.

3.3 Pre-market approvals

3.3.1 Pre-market approvals in the EU

The 1962 Directive concerning colours for use in foodstuffs,[21] which marked the beginning of European food law, was also the first European pre-market approval scheme. The directive harmonised Member States' legislation by establishing a single list of colouring matters whose use was authorised for colouring foodstuffs and laying down purity criteria for these colours. Positive lists are part of the law

---

[19] In accordance with the Statements of Principle Concerning the Role of Science in the Codex Decision-Making Process and the Extent to Which Other Factors Are Taken Into Account (CAC, 2011: Appendix)
[20] See Case C-77/09, *Gowan Comércio Internacional e Serviços Lda*, [2010] ECR 0000, paras 68-79.
[21] EEC: Council Directive on the approximation of the rules of the Member States concerning the colouring matters authorised for use in foodstuffs intended for human consumption, OJ 1962, 115/2645.

(usually as an annex). To include subsequently a product on the list (or delete it), the law must be changed by the applicable procedure.

While the details may differ, the system of positive lists and prior authorisation procedures for certain products plays an important role in the EU food safety policy. The pre-market approvals exist in European as well as in national food laws. Apart from novel foods, examples at EU level include food additives (incl. sweeteners, colourants, etc.)[22] flavourings,[23] extraction solvents,[24] infant formulae and follow-on formulae,[25] foodstuffs intended for particular nutritional uses,[26] food supplements,[27] genetically modified (GM) food and feed,[28] food contact materials,[29] and nutrition and health claims made on foods.[30]

3.3.2 Prior authorisation procedure for novel foods

Regulation 258/97 on novel foods and novel food ingredients (Novel Food Regulation – NFR)[31] applies to food that was not consumed to a significant degree in the EU before 15 May 1997, which is the date of entry into force of the Novel Food Regulation, and which falls under one of the following categories:
- its primary molecular structure is new or intentionally modified;
- it consists of or is isolated from microorganisms, fungi or algae;
- it consists of or is isolated from plants and food ingredients isolated from animals, except for foods and food ingredients obtained by traditional propagating or breeding practices and having a history of safe use;
- a production process which was applied to it is not currently used, where that process gives rise to significant changes in the composition or structure of foods or food ingredients which affect their nutritional value, metabolism or level of undesirable substances.[32]

---

[22] Reg. 1333/2008 on food additives, OJ 2008, L 354/16.
[23] Reg. 1334/2008 on flavourings, OJ 2008, L 354/34.
[24] Dir. 2009/32 on extraction solvents, OJ 2009, L 141/3.
[25] Dir. 2006/141 on infant formulae and follow-on formulae, OJ 2006, L 401/1.
[26] Dir. 2009/39 on foodstuffs intended for particular nutritional uses, OJ 2009, L 124/21.
[27] Dir. 2002/46 on food supplements, OJ 2002, L 183/51.
[28] Reg. 1829/2003 on genetically modified food and feed, OJ 2003, L 268/1.
[29] Reg. 1935/2004 on materials and articles intended to come into contact with food, OJ 2004, L 338/4.
[30] Reg. 1924/2006 on nutrition and health claims made on foods, OJ 2007, L 404/3.
[31] OJ 1997, L 43/1.
[32] Art. 1 NFR. In practice, the NFR concerns innovative foods (e.g. DHA-rich oil, phytosterols), foods produced by new technologies (e.g. high-pressure processing), as well as exotic traditional foods (e.g. noni fruit – *Morinda citrifolia* or *Stevia rebaudiana*), which have a history of safe use in other parts of the world, but which were not known in the EU prior to 1997. Since 2004, the pre-market approval procedure for GM foods has been regulated separately.

A manufacturer who wishes to place a novel food on the EU market must undergo a complicated authorisation procedure.[33] An application with detailed scientific information[34] supporting it is made to a single Member State. The Member State has 90 days to produce an initial assessment report and to decide whether an additional risk assessment is required (by EFSA). The initial assessment report is also circulated to all Member States, who have 60 days to comment or make a reasoned objection. If no additional risk assessment is required or if there are no objections from the Member States, a decision on the authorisation will be taken by the Member State where the application was submitted. Otherwise, the Commission takes an authorisation decision after EFSA has provided an opinion on the safety of the novel food.

Applying for an authorisation is a time-consuming and costly procedure imposing heavy burdens on manufacturers. It takes 35 months on average, but – in extreme cases – almost 10 years, for the Commission to decide on an authorisation. This wide time span results from the lack of time preset in the Novel Food Regulation. Such a long time is of course not conducive to innovation.[35] Around 7-10 applications are submitted per year under the Novel Food Regulation (EC 2008). It seems impossible to assess the real size of the novel food market because of its diversity (covering many different products), as well as confidentiality policies and intellectual property rights issues. However, given the market potential in Europe for novel foods and the high level of innovation in the food industry, this number has to be considered very low (Van der Meulen, 2009).[36]

Although a decided advantage of prior authorisation schemes is a case-by-case safety assessment of food products, the food safety legislator faces the challenge to strike a fair balance between the requirements of the protection of human health and the interests of the business sector and of the internal market. The

---

[33] The authorisation for novel food is specific – addressed to the applicant, so others cannot market the same food. The NFR, however, provides a simplified procedure for foods that are similar (the so-called substantial equivalence application) to a food that is already on the market.

[34] See Council Recommendation 97/618/EC, OJ 1997, L 253/1.

[35] Brookes estimates that the costs of bringing a novel food to market are between € 4 million to € 15.4 million, incl. R&D costs and the costs of meeting regulatory requirements, which vary between € 0.3 million and € 4 million. Costs related to the considerable additional time needed to obtain authorisation in the EU compared to other countries (similar legislation exists, i.e. in Canada, Australia and New Zealand; in the US a substance that will be added to food is subject to a pre-market safety assessment, unless its use is generally recognized as safe by qualified experts) add an extra € 0.3 million to € 0.75 million per application (Brookes, 2007).

[36] Although it is difficult to extrapolate information available on certain products, the world of exotic foods alone gives an idea on the potential for novel foods in the European market. The impact assessment study provides interesting numbers in this regard. World-wide, some 30 food plants supply 95% of the daily human intake of plant food calories. The remaining 5% in Europe come from around 300 other plant species, compared to nearly 7000 species used outside the EU (EC, 2008). Thus, from more than 6,600 existing food sources, fewer than 10 exotic plants have been approved since the NFR entered into force.

above observations beg the question whether the considerable obstacles the prior authorisation scheme places on businesses bringing novel foods to the market are justified.

3.4 Prior authorisation schemes versus risk analysis and the precautionary principle

3.4.1 Reversal of burden of scientific proof

The reason for prior authorisation procedures is to ensure a high level of health protection by requiring that some foods must undergo a scientific risk assessment to verify their safety prior to placing on the market. This constitutes an exception to the general rule that food enters the market without prior approval. These exceptions are introduced for categories of food that are considered *a priori* dangerous – until scientific evidence proves that a foodstuff is safe.

Prior authorisation schemes thus reverse the burden of providing scientific evidence. While, as a general rule, it is for society (public authorities, consumer organisations, citizens) to prove, on the basis of risk assessment, that a certain substance, product or technology is unsafe and has to be banned or restricted on the market, in the case of pre-market approvals the burden of proof is shifted onto businesses. A product is presumed to be dangerous until the proponent of placing it on the market provides scientific evidence stating otherwise. The applicants have a commercial interest in placing the product on the market and – therefore – in actively contributing to risk assessment procedures. Hence, the burden lies with the applicant to prove safety and not with others to prove harm (Levidow *et al.*, 2005).

The already mentioned EC Communication on the Precautionary Principle clearly refers to prior authorisation schemes (positive lists) as one of the possible expressions of the precautionary principle. According to this interpretation, the precautionary principle requires that responsibility for carrying out scientific work needed to assess the risk is shifted onto the business community. As long as there is no sufficient certainty that a product (or production method) is safe to human health, authorisation to place it on the market is refused (EC, 2000b: 21).

Thus, in the Commission's document, a clause shifting the burden of proof by requiring that certain substances are unsafe until proven otherwise is considered an example of the application of the precautionary principle. Precautionary food safety measures based on prior authorisation schemes are 'temporary' in the sense that certain products are not allowed, pending scientific data that would confirm their safety, and giving the opportunity to finance research necessary to carry out a risk assessment to those who have an economic interest in placing the products on the market.

3.4.2 Precaution: from policy guidelines to legal definition

The precautionary principle, as well as preventive and polluter pays principles, are not defined in the Treaty. The purpose of these principles is to set out general policy directions and to guide policy makers (De Sadeleer, 2002, 2010). It is for policy makers to flesh them out and make them operative though specific policy instruments.

The purpose of the Communication on the Precautionary Principle was to inform all interested parties of the manner in which the principle was interpreted and applied (EC, 2000b: 8). The document was adopted in 2000, two years before the General Food Law entered into force. At that time, the only explicit reference to the precautionary principle was in the environmental title of the Treaty (now Article 191(2) TFEU). The scope of application of the principle in EU law, however, was not limited to the environment, but also extended to human, animal and plant health. In the *Artegodan* case, the Court of First Instance (now renamed the General Court) confirmed that the precautionary principle has the status of an autonomous principle in EU law because the requirement that the protection of public health, safety and the environment applies to all EU policies.[37]

The Communication on the Precautionary Principle belongs to the category of 'atypical acts'. Atypical acts are not provided for by Article 288 of the Treaty on the Functioning of the European Union (TFEU) which sets out the secondary legislation of the EU. These not legally binding 'soft law' instruments, such as communications or guidelines, produce, however, important practical effects. According to settled case-law, EU institutions may lay down for themselves guidelines to exercise their discretionary powers by way of measures not provided in the Treaty (such as communications), provided that they contain directions on the approach to be followed by the EU institutions and do not depart from the rules of the Treaty. Courts verify whether a disputed measure is consistent with the guidelines laid down in such communications.[38] The European judiciary states, 'where the Commission adopts measures which are designed to specify the criteria which it intends to apply in the exercise of its discretion, it itself limits that discretion in that it must comply with the indicative rules ... imposed upon itself.'[39]

Based on the Treaty and – since 2000 – the Commission's Communication, which established further guidelines for the application of the precautionary principle, the principle has become normative in secondary legislation in a number of measures

---

[37] Joined Cases T-74/00, T-76/00, T-83/00, T-84/00, T-85/00, T-132/00 & T-141/00, *Artegodan and Others v. Commission*, [2002] ECR II-4945, paras 183-184.

[38] Case T-13/99, *Pfizer Animal Health SA v. Council*, [2002] ECR II-3305, para. 119 and case-law referred to therein.

[39] Joint Cases T-254/00, T-270/00 and T-277/00, *Hotel Cipriani SpA and Others v. Commission*, [2008] ECR II-3269, para. 292.

of a precautionary character aimed at ensuring a high level of the protection of human health. Not only measures taken in the field of food safety, such as the Novel Foods Regulation, but also other measures to protect human health under the common agricultural policy, e.g. legislation on plant protection products, provide examples of prior authorisation procedures based on the precautionary principle.

The principle also became gradually concretised in case-law, where, with regard to EU measures, the European judiciary followed the flexible, precaution-oriented approach outlined in the 2000 Communication, granting the EU institutions a wide discretion in decision-making and – as shown above – allowing them to stretch the precautionary principle to justify risk management measures departing from the findings of scientific risk assessments.[40]

If we accept that the European approach of placing the precautionary principle at the risk management stage allows the EU institutions more discretion, this discretion in the field of food safety has been crucially shaped by the EU risk analysis methodology and the definition of the precautionary principle introduced by the 2002 General Food Law. According to Article 4(3) GFL, all food law principles and procedures existing prior to the Regulation had to be adapted to the general principles of food law established in Articles 5-10 GFL by 2007. Moreover, existing legislation had to be implemented taking into account the new risk analysis framework and the definition of the precautionary principle from the beginning.[41]

Although prior authorisation schemes are commonly considered to be the application of the precautionary principle, shifting the burden of producing scientific evidence is not enough to call such procedures based on the precautionary principle (see also Chapter 4). Its definition in Article 7 GFL does not mention reversal of the burden of proof. Instead, it sets out pre-requisites for the application of the precautionary principle: measures must be preceded by scientific risk assessment which – although inconclusive – has identified some potentially dangerous effects of a product or process.

The relevance of the regulatory framework for risk analysis introduced by the General Food Law to food safety legislation and prior authorisation schemes was addressed for the first time by the European judiciary in 2010 in a judgement concerning national food safety measures. We will discuss it below.

---

[40] See, e.g. for prior authorisation schemes for plant protection products, Case C-77/09, *Gowan Comércio Internacional e Serviços Lda, supra* note 20.
[41] Art. 4(4) GFL.

3.4.3 National prior authorisation schemes put to the test

Prior authorisation requirements do not exist only at EU, but also at Member State level. Before the General Food Law entered into force, the European Court of Justice had the opportunity to address practices related to national pre-market approvals of fortified foods, i.e. foods to which vitamins or minerals have been added, in a number of rulings.[42] These judgements, however, did not touch upon national regulatory schemes as such, but referred to their implementation, *viz.* administrative practice entailing that foodstuffs enriched with vitamins or minerals could be marketed in that Member State only if it was shown that such enrichment with nutrients met a need of the population.[43]

In the *Commission* versus *France* judgement, the European Court of Justice had the opportunity – for the first time – to apply the precautionary principle and the risk analysis methodology set out in the General Food Law to food safety legislation.[44] The ruling concerned French measures laying down a prior authorisation requirement for processing aids and foodstuffs whose preparation involved processing aids.[45] This time, however, the action was directed at the prior authorisation scheme itself.

The Commission argued that recourse to a prior authorisation scheme, although not excluded in principle, should be targeted and precisely justified on a scientific basis.[46] The Court agreed with the Commission's opinion and confirmed that, in exercising their discretion relating to the protection of human health, the measures chosen by the Member States must be confined to what is necessary to attain this objective. More importantly, the Court stated in this regard:

> A Member State cannot justify a systematic and untargeted prior authorisation scheme ... by pleading the impossibility of carrying out more exhaustive prior examinations by reason of the considerable quantity of processing aids which may be used or by reason of the fact that manufacturing processes are constantly changing. As is apparent from Articles 6 and 7 of Regulation No 178/2002, concerning the analysis of risks and the application of the precautionary principle, such an approach does not correspond to the

---

[42] See Case 174/82, *Sandoz*, [1983] ECR 2445; Case 192/01, *Commission* v. *Denmark*, [2003] ECR I-9693; Case 95/01, *John Greeham and Léonard Abel*, [2004] ECR I-1333; Case C-24/00, *Commission* v. *France*, [2004] ECR I-1277; Case 41/02, *Commission* v. *Netherlands*, [2004] ECR I-11373; Case 270/02, *Commission* v. *Italy*, [2004] ECR I-1559.

[43] Case 41/02, *Commission* v. *Netherlands*, *supra* note 42, paras 22-23.

[44] Case 333/08, *Commission* v. *France*, [2010] ECR I-757.

[45] 'Processing aids' are substances intentionally used in the processing of raw materials, foods or their ingredients, to fulfil certain technological purposes during treatment or processing. They may result in the unintentional but technically unavoidable presence in the final product of residues of the substance, without any technological effect in the final product (Cf. Art. 3(2)(b) Reg. 1333/2008).

[46] Case 333/08, *Commission* v. *France*, *supra* note 44, para. 59.

requirements laid down in the Community legislature as regards both Community and national food legislation ...[47]

The French measure was judged disproportionate in relation to the possible risks which processing aids may pose for human health. For the first time in this context, however, the Court referred to the requirements of the risk analysis methodology established by the General Food Law, according to which food safety legislation normally has to be underpinned by an assessment of risks posed by a product or process.

The prior authorisation scheme for processing aids is an example of horizontal provisions. The scheme sets out a framework requiring all processing aids to undergo an authorisation procedure before they can be marketed on the French territory. Individual decisions concerning processing aids are the result of the implementation of this measure.

As a result of this analysis, the Court of Justice applied the risk analysis framework set out in the General Food Law to all food safety measures, regardless of whether they are legislative or regulatory acts. As mentioned above, the Regulation and the general principles contained therein apply to both EU and national levels. Turning to EU level and to the main question of this chapter, does the Court's interpretation of the General Food Law principles have any implications on the EU decision-making paradigm in this area of risk regulation?

3.4.4 Some are more equal than others?

Not only in national food safety laws does the absence of a scientific justification to consider certain foods a priori hazardous raise questions. It is also far from self-evident that this approach fits well with the underlying principles of food law when applied to EU legislation. But while national measures are ultimately put to the test of the free movement of goods principle, the internal market straitjacket does not seem to restrict the EU legislator to the same extent.

As illustrated above in the structure of EU food law, in light of the Treaty provisions, national food safety laws, which constitute a barrier to intra-EU trade, are exceptions to the fundamental principle of free movement of goods within the EU. These exceptions have to be interpreted strictly and cannot constitute a means of arbitrary discrimination or a disguised restriction of trade between Member States.[48] In practice, it means that Member States wanting to maintain or introduce their national food safety provisions that hinder EU trade have to prove that a substance

---

[47] *ibid.*, para. 103.
[48] Case 174/82, *Sandoz, supra* note 42, at 22; see also Case C-333/08, *Commission* v. *France, supra* note 44.

or product poses a genuine threat to public health.[49] In the context of the general principles of food law, they will need to provide scientific evidence (risk assessment) to justify their exceptions to free movement of goods.

Bound by the Treaty rules relating to the functioning of the internal market in the same way as the Member States, the EU institutions enjoy, however, a much broader discretion in taking measures. Unlike national food safety provisions, which – as exceptions to the free movement of goods principle – have to be interpreted strictly, EU measures do not need to be underpinned by an assessment of risk proving that they are absolutely necessary to protect human health. An example of the discretion applied to EU rules by Courts can be found in the *Fedesa* case, where the applicants argued that the EU directive banning the use of hormones in meat was not supported by scientific evidence and hence unlawfully hindered free movement of goods within the EU.[50] The Court did not uphold this claim, referring to the discretion conferred on the EU institutions, and stating that the Council was free to decide on adopting a solution that responded to the concerns expressed by society.[51]

Because the general principles of food law set out in Regulation 178/2002 form a framework of a horizontal nature applicable to all measures, the way the risk analysis methodology is applied to national legislation should – in principle – apply to EU food safety legislation as well. The launch of the EU model of risk analysis for the whole EU multi-level food safety governance and the 2007 deadline for adapting food safety legislation to the new principles did not result in profound reforms of the existing legislation at national or EU levels (Van der Meulen, 2006a,b). However, whereas Member States' food safety laws are submitted to the Treaty regime concerning free movement of goods, ensuring consistency of EU food safety legislation with the principles contained in the General Food Law would require some regulatory discipline, at least in relation to giving reasons for the regulatory choices of the EU legislator.

Consider the 2008 Proposal revising and updating the Novel Foods Regulation,[52] which eventually failed after the Council and European Parliament were unable to reach agreement on cloning in food production. Nevertheless, the proposal is an example of a measure that was designed after the General Food Law entered into force. Before 2008, no steps were taken to review the Novel Foods Regulation to adapt it to the General Food Law. The draft report on impact assessment for the proposal mentions Regulation 178/2002 among the EU legislation that had been taken into account in revising the Novel Food Regulation merely stating that the

---

[49] See Case C-192/01, *Commission v. Denmark, supra* note 42.
[50] Case C-331/88, *Fedesa*, [1990] ECR I-4023.
[51] *Ibid.*, para. 9.
[52] COM (2007) 872 final.

General Food Law lays down only general principles and requirements regarding food safety and thus does not address specific issues such as the pre-market safety assessment of food which is covered by sectoral legislation.

What makes the Novel Foods Regulation an interesting example is that the risk analysis methodology has unquestionably been applied within this framework: regulatory decisions concerning individual products are underpinned by a scientific risk assessment, and EFSA not only plays an active role in these procedures, but it also organises meetings to discuss scientific information needed for such applications (EFSA, 2010). Never in this context, however, has the rationale behind the framework itself been discussed. The Novel Foods Regulation applies to a wide range of different foods considered dangerous until proven safe, but no risk assessment underlies this presumption of lack of safety. No indication is available that EFSA's opinion had been asked prior to submitting the proposal or that it was based in some other way on risk assessment.

In the explanatory memorandum on the Proposal for a novel food regulation the Commission concluded that 'there was no need for external expertise'.[53] The fact that scientific expertise was not considered to be an important element in the design of the new novel food regulation confirms that the risk analysis methodology is not among the principles the European legislator is guided (limited) by in adopting food safety measures. This broad discretion, however, leads to a question of the conformity of EU legislation with international trade rules defined by the WTO agreements.

3.4.5 WTO regime – external yardstick

Although the choice of an appropriate level of protection is perceived as a democratic choice of each WTO member, measures applied to protect human, animal or plant health must meet rather strict risk assessment requirements to be considered justified barriers to trade. Because marketing approval for novel foods is conditional and depends, i.e. on a satisfactory demonstration that the product does not present a danger to human health, the Novel Foods Regulation

---

[53] *Ibid.*: 4.

falls under the scope of the definition of an 'SPS measure',[54] which means that it is governed by the SPS Agreement.[55]

In general, prior authorisation schemes are not forbidden under WTO law, provided that certain conditions are met. Article 8 of the SPS Agreement permits procedures aimed at 'checking and ensuring the fulfilment of SPS measures' and undertaken in the context of 'control, inspection, or approval'.[56] Annex C sets out requirements for these procedures. They have to be, i.e. undertaken and completed without undue delay and in no less favourable manner for imported than for like domestic products; the applicant has to be duly informed by the competent authorities about the progress of the application at all stages of the procedure; and any requirements for control, inspection and approval of individual specimens of a product must be limited to what is reasonable and necessary.

Hence, within the meaning of these provisions, approvals are part of the procedures applied to implement an SPS measure.[57] The SPS measure is in this case the regulatory framework establishing prior authorisation requirements. In the *Biotech products* case, the WTO panel had the opportunity to consider the risk assessment requirement in the context of individual authorisation decisions taken within the framework of the Novel Foods Regulation, but did not consider the regulatory framework itself.[58] However, subsequent complaints over the Novel Foods Regulation raised by some developing countries, including Peru, Ecuador and Colombia, refer to the prior authorisation scheme as such, and it is likely that, this time, the prior authorisation scheme itself will be brought to the WTO Dispute Settlement Mechanism.

---

[54] Annex A(1) reads: '*Sanitary or phytosanitary measure* – Any measure applied: (a) to protect animal or plant life or health within the territory of the Member from risks arising from the entry, establishment or spread of pests, diseases, disease-carrying organisms or disease-causing organisms; (b) to protect human or animal life or health within the territory of the Member from risks arising from additives, contaminants, toxins or disease-causing organisms in foods, beverages or feedstuffs; (c) to protect human life or health within the territory of the Member from risks arising from diseases carried by animals, plants or products thereof, or from the entry, establishment or spread of pests; or (d) to prevent or limit other damage within the territory of the Member from the entry, establishment or spread of pests. Sanitary or phytosanitary measures include all relevant laws, decrees, regulations, requirements and procedures including, *inter alia*, end product criteria; processes and production methods; testing, inspection, certification and approval procedures; quarantine treatments including relevant requirements associated with the transport of animals or plants, or with the materials necessary for their survival during transport; provisions on relevant statistical methods, sampling procedures and methods of risk assessment; and packaging and labelling requirements directly related to food safety.'

[55] This conclusion was reached in WTO Reports of the Panels, European Communities – Measures Affecting the Approval and Marketing of Biotech Products, WT/DS291/R; WT/DS292/R; WT/DS293/R, adopted 29 Sept. 2006, para. 7.427.

[56] *Ibid.*, para. 7.424.

[57] *Ibid.*, para. 7.1491.

[58] *Ibid.*, paras 1525-1526. Until April 2004, the scope of the NFR included GM foods.

The concerns were first raised in 2006 by Peru, which highlighted in its communication that, as a consequence of the implementation of the Novel Foods Regulation, exports of exotic traditional plants to the EU had been stopped.[59] The main objection was the lack of distinction between strictly novel foods, i.e. those that have not been consumed anywhere in the world, and those that are novel only in the EU, e.g. exotic traditional products with a history of safe use outside the EU. Such products are submitted to the same prior authorization procedure, in which the applicant has to prove that a product is safe to consumers. These safety considerations refer to a category of products determined solely on the basis of an arbitrary date (15 May 1997), despite the fact that some of them have been used safely for human consumption for centuries in the country of origin and elsewhere in the world.

The complaint against the EU Novel Foods Regulation refers to the same arguments as those raised against the French prior authorisation schemes for processing aids. The Novel Foods Regulation, by applying a prior authorisation scheme to a wide range of products without scientific justification, is considered an untargeted and arbitrary measure, disproportionate in relation to possible risks which products falling under the scope of this legislation may pose to human health. In consequence, the intervening countries claim that the Regulation creates an unnecessary and unjustified barrier to international trade because of the very high costs of producing the scientific studies required and a lengthy authorisation procedure.[60]

According to the SPS Agreement, WTO members have the right to adopt sanitary or phytosanitary measures necessary for the protection of human, animal or plant life or health, provided that such measures are based on scientific principles and not maintained without sufficient scientific evidence.[61] Hence, when a WTO member invokes the precautionary principle to justify its measures, the exception set out in Article 5.7 SPS is the standard by which the measures will be judged whether they are justified and necessary. Because the Appellate Body declines to recognize the precautionary principle as a general principle of international law which could override obligations under the SPS Agreement,[62] Article 5.7 SPS considerably limits the inclinations of the EU legislator to dress up measures based on factors other than science in the clothing of 'precaution-oriented' legislation.

Although a detailed discussion on the effects of WTO law and DSB rulings in the EU legal order is outside the scope of this chapter, it has to be noted that – generally

---

[59] G/SPS/GEN/681 (5 Apr. 2006).

[60] G/SPS/GEN/713 (12 July 2006). The trade concerns regarding Reg. 258/97 were raised again in 2011, after the EU institutions failed to agree on the revision of the Regulation. See G/SPS/GEN/1087 (7 June 2011).

[61] Art. 2 SPS.

[62] WTO Appellate Body Report, *EC-Hormones, supra* note 17, para. 123.

– unlike other agreements concluded by the European Union, the EU Courts do not consider the WTO Agreements among the rules in light of which the legality of measures adopted by the EU is reviewed.[63] This stance has triggered much criticism in literature (Griller, 2000; Mendez, 2004; Snyder, 2003; Zonnekeyn, 2004).

Two important exceptions to this line of jurisprudence have been recognised by EU Courts. GATT/WTO provisions have the effect of binding the EU where the EU implements a particular obligation (*Nakajima* exception),[64] or where an EU measure refers expressly to specific GATT/WTO provisions (*Fediol* exception).[65] Although the way the EU Courts have so far interpreted these exceptions is considered to be rather narrow and rarely applied (Zonnekeyn, 2001), the *Nakajima/Fediol* doctrine might still have important implications for food safety regulation. EU measures in this field often explicitly refer to the SPS Agreement, so they should, in principle, meet this standard. An example of the WTO 'consciousness' is reflected, e.g. in the EU Regulation on the hygiene of foodstuffs. Recital 18 states that the Regulation takes account of international obligations laid down in the SPS Agreement and the international food safety standards contained in the *Codex Alimentarius*.[66]

3.5 Conclusion

This chapter illustrates the tensions between the discretion of the EU legislator and the demand for science-based risk regulation. Food safety is the only area of risk regulation where a comprehensive risk analysis model has been introduced not only by international trade agreements, but also by EU legislation, as one of the general principles governing the food safety policy, applied to both EU and national measures.

The ECJ ruling on the French measures concerning processing aids and the analysis of the EU Novel Foods Regulation in light of the WTO SPS Agreement show that the risk analysis methodology relates not only to product authorisations and other technical 'decisions under legislation', but also to the choice and design of horizontal, framework legislation.

Under both WTO and EU laws, all food safety measures that restrict trade must be science-based and recourse to the widely interpreted and flexible concept of precaution by the EU legislator risks non-compliance with the decision-making

---

[63] The final recital of Dec. 94/800/EC concerning the conclusion on behalf of the European Community, as regards matters within its competence, of the agreements reached in the Uruguay Round multilateral negotiations (1986-1994) explicitly denies the direct applicability of WTO rules (OJ 2004, L 336/1).
[64] Case C-69/89, *Nakajima All Precision Co. Ltd v. Council*, [1991] ECR-2069, para. 31.
[65] Case 70/87, *Fédération de l'industrie de l'huilerie de la CEE (Fediol) v. Commission*, [1989] ECR 1781, paras 19-22.
[66] Reg. 852/2004, OJ 2004, L 226/3.

paradigm in this field, if precautionary measures are based on considerations other than science. In this sense, risk analysis defines the boundaries of EU food safety legislation and limits democratic choices for stricter regulations.

4. The impact of the definition of the precautionary principle in EU food law[*]

Abstract

Regulation 178/2002 contains a definition of the precautionary principle. This is the first time a legal definition of the principle has been formulated for all EU food law. This fact, however, has been given hardly any attention in literature. It is all the more surprising because the lack of clear formulation of this concept was the main reason for criticism and controversy surrounding the principle. In the absence of a definition, the precautionary principle was a general notion, similar to the principles of subsidiarity or proportionality. Its scope was determined by decision makers and ultimately shaped by tendencies in case law. After codification, existing policy orientations and earlier case law may in some aspects differ from the legal definition. This chapter characterizes the main constituents of the precautionary principle and the risk analysis methodology established in Regulation 178/2002 and discusses examples illustrating how different provisions of the Regulation narrow the scope of the principle. It focuses on issues critical to its application: emergency measures, the so-called other legitimate factors, the burden of proof, and pre-market approval schemes.

4.1 Introduction

Regulation 178/2002 laying down the general principles and requirements of food law, establishing the European Food Safety Authority and laying down procedures in matters of food safety[1] (the so-called General Food Law – hereinafter referred to as GFL or the Regulation) contains a definition of the precautionary principle. This is the first time EU law has formulated a legal definition of the principle. Previously, its application to different areas was based on the Treaty, which only adopts the principle, but without determining either conditions of putting it into practice or consequences of its implementation.

The definition of the precautionary principle in EU food safety law constitutes an important step in helping to determine the characteristics of the principle applied in this area. It is all the more important because the lack of a clear formulation of the concept and its numerous different interpretations were the main reasons for criticism of the principle (Foster *et al.*, 2000).

---

[*] Reprinted from Common Market Law Review, 47, Szajkowska, A., The impact of the definition of the precautionary principle in EU food law, 173-196, 2010, with permission from Kluwer Law International.

---

[1] OJ 2002, L 31/1.

In the absence of a definition, the precautionary principle was a general notion, similar to the principles of subsidiarity or proportionality. Its scope was determined by the decision makers and ultimately shaped by tendencies in case law (EC, 2000b: 10). After codification, existing policy orientations and case law concerning the precautionary principle may in some aspects differ from the definition in the legislation. Moreover, the General Food Law has also introduced a comprehensive model for risk analysis. According to the Regulation, the precautionary principle is part of the risk analysis methodology. Therefore, it has to be interpreted strictly within this context (Belvèze, 2003).

The precautionary principle is certainly not a new concept, and a great deal has been written on this subject. However, the fact that the principle has been given a legal definition in the area of food safety has received hardly any attention in literature. Rather, in the research focusing on the precautionary principle in EU law, the concept is evaluated as a regulatory continuum (Christoforou, 2003; Dutheil de la Rochère, 2007; Fisher, 2009), or is thought of as a general policy guidance (Majone, 2002).

Thus, this chapter focuses on the cases that illustrate how different provisions of the General Food Law may narrow the scope of the precautionary principle applied in the field of food safety. It starts with a short overview of the concept of the precautionary principle, referring to the EU Treaties, legislation and case law. Then, it analyses the definition of the precautionary principle in the Regulation and characterizes its main constituents by referring to the broader context of the risk analysis methodology established therein. It explores the consequences of the definition of the principle if we presume that this definition constitutes an all-inclusive, exhaustive codification in the field of food safety. Three examples – two from case law prior to the General Food Law and a regulatory framework for genetically modified food and feed – are put to the test of the conditions set out in the new definition. The chapter discusses the impact of the definition of the precautionary principle in the light of the following critical issues: emergency measures, the so-called other legitimate factors in risk management, the burden of proof and pre-market approval schemes.

4.2 Evolution of precaution: from 'scientific uncertainty' to the 'precautionary principle'

4.2.1 Precautionary principle in EC Treaty and TFEU

Although recognition of the precautionary principle in different areas of risk regulation had been increasing gradually since its inception in German environmental policy in the late 1970s, it took more than twenty years until the principle appeared for the first time in EU law. Even the predecessor of Article 191 TFEU (ex Art. 174 EC), Article 130r added by the Single European Act in

1986, did not mention the precautionary principle among the principles of EC environmental policy. The first explicit reference to the principle was introduced by the Maastricht Treaty. Article 191(2) TFEU (ex 174(2) EC) provides:

> Union policy on the environment shall aim at a high level of protection taking into account the diversity of situations in the various regions of the Union. It shall be based on the precautionary principle and on the principles that preventive action should be taken, that environmental damage should as a priority be rectified at source and that the polluter should pay.

The scope of application of the precautionary principle in EU law has not been limited to the environment, however, but also includes human, animal and plant health. This broad interpretation derives directly from Article 191(1) TFEU, which states that Union policy on the environment shall contribute to pursue the objective of protecting human health, and from Article 11 TFEU (ex Art. 6 EC), which stipulates that environmental protection requirements must be integrated into the definition and implementation of Union policies and activities.

In view of this, the Court of First Instance confirmed in *Artegodan* that, because the requirement of the protection of public health, safety and the environment applies to all spheres of Union activity, the precautionary principle has the status of an autonomous principle in EU law.[2] Hence, the EU institutions are obliged to consider the precautionary principle in their policies to assure a high level of environmental and health protection.

4.2.2 Case law before the General Food Law

In the absence of a definition of the precautionary principle in the EC Treaty (or the TFEU), the European Court of Justice and the Court of First Instance (now renamed the General Court) attempted to explain the concept. An overview below concentrates on some of the most important judgments referring to the concept of the precautionary principle in the field of health protection and food safety.

The core element of the GFL definition of the precautionary principle – decision making under scientific uncertainty and a margin of appreciation associated with it – already emerged in rulings from the Luxembourg judiciary in the 1980s, even before the term 'precautionary principle' was introduced in the EC Treaty. In the *Sandoz* case concerning national rules prohibiting without prior authorization the marketing of foodstuffs to which vitamins have been added, the Court stated, 'In so far as there are uncertainties at the present state of scientific research, it is for the Member States, in the absence of harmonization, to decide what degree

---

[2] Joined Cases T-74/00, T-76/00, T-83/00, T-84/00, T-85/00, T-132/00, T-137/00 & T-141/00, *Artegodan and Others* v. *Commission*, [2002] ECR II-4945, at 183-184.

of protection of the health and life of humans they intend to assure, having regard however for the requirements of the free movement of goods within the Community'.[3]

In the judgments on the validity of the EU measure imposing an embargo on British beef exports to deal with the outbreak of 'mad cow disease' (bovine spongiform encephalopathy – BSE),[4] the Court of Justice formulated an approach that was repeated in subsequent rulings, and that was widely recognized as based on the precautionary principle. Although the transmissibility of BSE to humans at that time could not be confirmed, new scientific information did not rule out a possible link between BSE and Creutzfeldt-Jakob disease. In reply to growing consumer concern over safety of beef and the possibility of a serious public health hazard, the Commission adopted an emergency measure banning all exports of British beef.[5] The ban was temporary, subject to review based on new scientific information allowing for a more comprehensive analysis of the situation. Against this backdrop, the Court did not find the measure to be in breach of the principle of proportionality, and upheld the legality of the decision, justifying its judgment as follows: 'Where there is uncertainty as to the existence or extent of risks to human health, the institutions may take protective measures without having to wait until the reality and seriousness of those risks become fully apparent'.[6]

Although the *BSE* judgment did not refer in so many words to the precautionary principle, it is considered one of the most important early expressions of this concept. The first explicit reference to the precautionary principle that can be found in a Court judgment is in Case 6/99, *Greenpeace France* v. *Ministère de l'Agriculture et de la Pêche*, concerning deliberate release of genetically modified organisms into the environment.[7] However, this case does not seem to analyse scientific uncertainty so much as the possible emergence of new information regarding risks.

Finally, in the *Pfizer* case, the Court of First Instance took the opportunity to examine in more detail the main elements of the precautionary principle. In this case the notion of scientific uncertainty as elaborated in the *BSE* case and the label precautionary principle were brought together: 'Where there is scientific uncertainty as to the existence or extent of risks to human health, the Community

---

[3] Case 174/82, *Sandoz*, [1983] ECR 2445, at 16.

[4] Cases C-157/96, *National Farmers' Union*, [1998] ECR I-2211; and 180/96, *United Kingdom* v. *Commission*, [1998] ECR I-2265.

[5] Dec. 96/239/EC of 27 March 1996 on emergency measures to protect against bovine spongiform encephalopathy, OJ 1996, L 78/47.

[6] *National Farmers' Union*, *supra* note 4, at 63; and *United Kingdom* v. *Commission*, *supra* note 4, at 99.

[7] Case 6/99, *Greenpeace France* v. *Ministère de l'Agriculture et de la Pêche*, [2000] ECR I-1651. However, it must be noted that the term 'precautionary principle' appeared earlier in Opinions of advocates general. See, e.g. Opinion of A.G. Gulmann of 28 Sept. 1993 in Case 405/92, *Etablissements Armand Mondiet SA* v. *Armement Islais SARL*, [1993] ECR-6133.

institutions may, by reason of the precautionary principle, take protective measures without having to wait until the reality and seriousness of those risks become fully apparent'.[8]

4.2.3 Precautionary principle in international and EC food safety regulations

The European Union faced the BSE outbreak shortly after the precautionary principle had been included in the environment title of the EC Treaty. The outbreak, viewed by Franz Fischler, then Commissioner for Agriculture, as the biggest crisis in the history of the EU (Ratzan, 1998), coincided in time with the conclusion of the World Trade Organization (WTO) Agreement on Sanitary and Phytosanitary Measures (SPS). These events not only shaped the discussion on the precautionary principle, but were also a catalyst for major changes in the EU regulatory framework for food safety.

The strict scientific regime of the SPS Agreement provides an exception that allows its members to adopt (provisionally) a health protection measure that constitutes a barrier to international trade in cases where scientific uncertainty prevails.[9] In this context, the possibility to break free of the obligation to provide scientific evidence becomes a powerful trade policy tool. No wonder the controversies around the definition of the precautionary principle had already arisen in the first case to interpret the SPS Agreement – a beef hormone dispute brought against the EU by the US and Canada.[10] The principle also stirred heated debate in the Codex Alimentarius Commission (CAC, 1999), where, generally, the United States defended the view that precaution is inherent in risk assessment, and that linking it to risk management – as advocated in the EU approach – would enable decision makers to overrule scientific risk assessment.[11]

By 1997, the Commission had already set out general policy guidelines in the Green Paper on the General Principles of Food Safety (EC, 1997a) and in the

---

[8] Case T-13/99, *Pfizer Animal Health SA* v. *Council*, [2002] ECR II-3305. See also Case T-70/99, *Alpharma* v. *Council*, [2002] ECR II-3495. For comments on the *Pfizer* judgment see De Sadeleer (2006) and Da Cruz Vilaça (2004).

[9] Art. 5.7 SPS.

[10] WTO Appellate Body Report, EC – Measures Concerning Meat and Meat Products, WT/DS26/AB/R; WT/DS48/AB/R, adopted 13 Feb. 1998.

[11] With reference to the application of precaution within Codex, the US 'expressed its objection to the inclusion of the precautionary principle as there was no internationally recognized definition and a precautionary approach was already built in risk assessment; this concept should not be used by risk managers to overrule risk assessment. The Delegation recalled that under Article 5.7 of the SPS Agreement, national governments may adopt provisional measures in cases of insufficient scientific evidence but they should seek to obtain additional information for a more objective assessment of risk; at the international level and in the framework of Codex, standards should be based on scientific evidence' (CAC, 1999). For differences in American and European risk regulations see also: Löfstedt and Vogel (2001), Post (2006), Vogel (2001), and Wiener and Rogers (2002).

Communication on Consumer Health and Food Safety (EC, 1997b). Three years later, the EC issued the White Paper on Food Safety (EC, 2000a). In all these documents, the precautionary principle was considered a key tenet of the new policy. The reform was designed to attain a high level of human health protection and was built on a precautionary approach as 'the rule in case scientific evidence is incomplete or unconvincing one way or the other' (EC, 1997a: 10).

It was also around that time that the term 'precautionary principle' started to appear in EU food safety legislation, from the beginning finding concrete expression. Recital 4 of Directive 1999/39/EC amending Directive 96/5/EC on processed cereal-based foods and baby foods for infants and young children reads:

> [W]hereas, taking into account the Community's international obligations, in cases where the relevant scientific evidence is insufficient, the precautionary principle allows the Community to provisionally adopt measures on the basis of available pertinent information, pending an additional assessment of risk and a review of the measure within a reasonable period of time.[12]

In 2000 the European Commission published the Communication on the Precautionary Principle to set out its position on application of the principle and to confirm once again that the principle is critically important for EU food safety policy. The Commission's stance was endorsed by the December 2000 Nice European Council Resolution, which called on the Commission 'to incorporate the precautionary principle, wherever necessary, in drawing up its legislative proposals and in all its actions' (European Council, 2000a). Even though the Communication does not have a legally binding force, it does have important practical effects: EU courts have made numerous references to this document as reflecting 'the law as it stood at that time'.[13] The Communication was generally perceived as a response to criticism of the EU policy from international *fora*, although the Commission itself asserted that it had published the Communication for purely EU reasons: to ensure a harmonized approach of the Member States towards the BSE crisis (Van der Haegen, 2003).

Finally, in 2002, the General Food Law entered into force, introducing the first legal definition of the precautionary principle in EU law.[14] Although the development of the concept of the precautionary principle seems to be more dynamic and advanced in the field of public health – and food safety in particular – up to now no universally accepted interpretation of the precautionary principle has emerged at the international level. In the discussion over the precautionary principle that reappeared in the WTO dispute settlement system in the *Biotech Products* case

---

[12] OJ 1999, L 124/8.
[13] *Pfizer, supra* note 8, at 123.
[14] Regulation 178/2002, *supra* note 1.

brought against the EC by the US, Canada and Argentina,[15] the Panel refrained from formulating a definite view on the question whether the precautionary principle constitutes a recognized principle of general or customary international law. It also added that, since the *Hormones* ruling, no authoritative decision by an international court or tribunal had recognized the precautionary principle as a general principle in international law, and no precise definition and content of the precautionary principle existed. Therefore, its status remained unsettled.[16]

4.3 Precautionary principle in the General Food Law

4.3.1 Science-based measures

The principles laid down in the General Food Law form a general framework for the new EU policy on food safety. The principle of risk analysis set out in Article 6 GFL introduces a requirement that food law shall be science-based. The precautionary principle is codified in Article 7 GFL, and has been placed within the structured approach to risk analysis.

In risk regulation, where most standards heavily depend on scientific data, risk analysis provides decision makers with a systematic approach to make decisions based on the best available scientific knowledge. Von Moltke describes the risk analysis methodology as 'bridging the gap between science and policy', and finding 'specific justification for each critical decision' (Von Moltke, 1996). In short, regulatory actions based on risk analysis are a two-step process. First, in a risk assessment procedure, experts determine the health effects of exposure of individuals or a population to hazardous materials or situations (NRC, 1983: 3). Risk management is carried out by politicians, who decide what hazards present more danger than society is willing to accept. It is a process of considering policy alternatives and selecting the best regulatory action to reduce or eliminate risks, taking into account risk assessment as the basic output, but also wider social values, economic, political concerns, and other dimensions of the problem, to which science alone cannot provide all answers.

In line with the division between scientific risk assessment and political risk management, some authors distinguish the 'precautionary approach' as an established part of risk assessment from the 'precautionary principle' defined as an entirely separate risk management concept (De Sadeleer, 2002; Graham and Hsia, 2002). Others identify 'caution' applied by scientists in risk assessment, and 'precaution' applied to the risk management stage solely (Levidow *et al.*, 2005). Although at first glance the issue seems to be rather taxonomic and of

---

[15] WTO Reports of the Panels, European Communities – Measures Affecting the Approval and Marketing of Biotech Products, WT/DS291/R; WT/DS292/R; WT/DS293/R, adopted 29 Sept. 2006.
[16] *Ibid.*, 339.

little relevance to the legal debate (Da Cruz Vilaça, 2004; De Sadeleer, 2002), the question whether precaution applies to risk management or risk assessment is in fact – as shown in the transatlantic conflict over the precautionary principle – a crucial point in the international discussion and the key to understanding the controversy over the principle.

Precaution built in risk assessment, forcing scientists to identify uncertainties and calling for more scientific evaluation to eliminate them, is different from the precautionary principle applied in risk management. The principle as a risk management tool leaves much more appreciation for decision makers, giving them the power to arbitrate whether scientific information is sufficient and certain, and to bias uncertainties in favour of safety, even against the main conclusions of scientific risk assessment.[17] The Commission in its Communication also makes a distinction between 'prudential approach' and 'precautionary principle', considering the former to be 'an integral part of the scientific opinion delivered by the risk evaluators', and the latter to be part of risk management (EC, 2000b: 13). The European legislature, however, made clear from the outset that the precautionary principle is above all a risk management tool.

Interestingly, although Article 6(3) GFL states that risk management should 'take into account' the results of scientific risk assessment, other factors legitimate to the matter under consideration, and the precautionary principle, the principle is not included in the definition of risk management itself. According to Article 3(12) GFL, risk management is 'the process, distinct from risk assessment, of weighing policy alternatives in consultation with interested parties, considering risk assessment and other legitimate factors, and, if need be, selecting appropriate prevention and control options'. The rationale for omitting the precautionary principle in the definition of risk management could be explained by the fact that the principle provides a *mechanism* within the risk analysis methodology, enabling decision makers to take a measure where the relevant scientific evidence is insufficient, whereas scientific risk assessment and other legitimate factors determine the *content* of the measure. In other words, the precautionary principle allows decision makers to take a decision under the conditions of scientific uncertainty, but it does not determine the decision itself.

The precautionary principle in the General Food Law, as part of the risk analysis methodology, is a directing principle addressed to public authorities. In the EU multi-level food safety governance, the principle applies to measures adopted both at EU and national levels. Recital 16 GFL states that '[m]easures adopted by the Member States and the Community should generally be based on risk analysis' and that '[r]ecourse to a risk analysis prior to the adoption of such measures should facilitate the avoidance of unjustified barriers to the free movement of foodstuffs'.

---

[17] See e.g. Case T-392/02, *Solvay Pharmaceuticals BV* v. *Council*, [2003] ECR II-4555.

Hence, because one of the main objectives of risk analysis and the precautionary principle is to ensure a greater consistency in the decision-making process, they should be applied at all levels of EU governance. This view was confirmed by the European Council in its Resolution on the Precautionary Principle: '[T]he precautionary principle applies to the policies and action of the Community and its Member States and concerns action by public authorities both at the level of the Community institutions and that of Member States' (European Council, 2000b).

The principles set out in the General Food Law are of a horizontal nature and apply to 'all stages of production, processing and distribution of food, and also of feed produced for, or fed to, food-producing animals'[18] (the 'farm to fork' continuum), whereas 'food law' is defined as 'laws, regulations and administrative provisions, whether at Community or national level'.[19] Given that only primary production for private domestic use or the domestic preparation, handling or storage of food for private domestic consumption are exempted from the scope of application of the General Food Law, it must be concluded that the principles apply to virtually all food safety legislation, both at EU and national levels. In other words, the definition in Article 7 has to be followed every time public authorities invoke the precautionary principle in the field of food safety in the EU and Member States.

4.3.2 Definition of the precautionary principle

Article 7 GFL defines the precautionary principle and sets out pre-requisites that decision makers have to meet in putting the principle into practice (Textbox 4.1). According to the definition, three conditions have to be satisfied if the precautionary principle is to be considered:
1. available scientific data has to be assessed before a decision is made;
2. potentially dangerous effects deriving from a given phenomenon, product or process, have to have been identified;
3. the scientific evidence is inconsistent or inconclusive and does not allow the risk to be determined with sufficient certainty.

Precautionary measures may only be provisional, which means that they should be reviewed and assessed periodically in the light of scientific progress. Paragraph 2 of the Article states that the measures must be proportionate, reviewed within a reasonable period of time, and no more restrictive of trade than is required to achieve the high level of health protection.

In the following sections, we will refer to three examples of the application of the precautionary principle in the area of food safety and confront them with the definition of the principle.

---

[18] Art. 1(3) GFL.
[19] Art. 3(1) GFL.

Textbox 4.1. General Food Law Article 7: precautionary principle.

Article 7

Precautionary principle

1. In specific circumstances where, following an assessment of available information, the possibility of harmful effects on health is identified but scientific uncertainty persists, provisional risk management measures necessary to ensure the high level of health protection chosen in the Community may be adopted, pending further scientific information for a more comprehensive risk assessment.

2. Measures adopted on the basis of paragraph 1 shall be proportionate and no more restrictive of trade than is required to achieve the high level of health protection chosen in the Community, regard being had to technical and economic feasibility and other factors regarded as legitimate in the matter under consideration. The measures shall be reviewed within a reasonable period of time, depending on the nature of the risk to life or health identified and the type of scientific information needed to clarify the scientific uncertainty and to conduct a more comprehensive risk assessment.

4.4 Application of the precautionary principle to emergency measures

4.4.1 Rapid Alert System for Food and Feed: the Malagutti judgment

The first example deals with emergency measures taken in reaction to a message circulated via the Rapid Alert System for Food and Feed (RASFF). RASFF is a tool for exchanging information on human health risks deriving from food or feed. The network involves the Member States, the European Food Safety Authority (EFSA), the Commission, which is responsible for managing the network, and some third countries. The provisions governing the RASFF are set out in the General Food Law,[20] but the system of food safety information exchange has been operating since 1979.

The *Malagutti* judgment concerns compensation for damages suffered by a company after the Commission issued a rapid alert message notifying the presence of pesticide residues in apples from France.[21] The Icelandic contact point informed the Commission that a batch of apples of French origin distributed via the Netherlands was withdrawn from the market, after discovering a level of 0.8 mg/kg of dicofol in those apples. EU rules set the maximum dicofol level for apples at 0.02 mg/kg. Thus, the apples in question were believed to contain

---

[20] Arts. 50-52 GFL.
[21] Case T-177/02, *Malagutti-Vezinhet SA v. Commission*, [2004] ECR II-827.

dicofol at a level 40 times greater than the maximum permitted level. The alleged supplier in France was the company Malagutti. The Commission forwarded the message received from the Icelandic contact point to the other members of the network. Subsequently, all trade in apples from the company was halted in the UK, consignments were returned to France, and Malagutti had to pay the costs. Nine days after the notification, French authorities took samples at Malagutti's warehouse from the same category of apples as those disposed of in Iceland. No dicofol was detected. Malagutti also informed the Commission that it had never exported apples to Iceland and decided to take legal action to claim compensation for the damage suffered following the circulation of the messages under the RASFF.

4.4.2 Facets of uncertainty

The *Malagutti* judgment concerns an emergency situation in which effective and fast communication is a crucial aspect of the whole food safety system. The Court found that as far as the warning of risks for consumer health was concerned, 'plausible evidence of a link between the applicant and the apples objected to in Iceland' constituted sufficient information for the Commission to transmit the alert.[22] Interestingly, the Court referred to the precautionary principle to justify the Commission's action, stating:

> If any doubts remained in that regard [the link between the company Malagutti and the apples found in Iceland], it should be pointed out that under the precautionary principle prevailing in the matter of the protection of public health the competent authority may be obliged to take appropriate measures to prevent certain potential risks for public health without having to wait until the existence and seriousness of those risks has been fully demonstrated.[23]

Thus, the Court applied the precautionary principle, which allows adopting measures before information is complete, to the procedure of rapid information exchange. Would the situation described above justify recourse to the precautionary principle in light of the definition of the principle in the General Food Law? In *Malagutti*, authorities at the national and EU levels assessed the available information and identified the possibility of harmful effects. On this basis, an alert notification – a risk management measure – was issued without waiting until the remaining uncertainties were clarified. These facts indeed bear several similarities to a situation triggering the precautionary principle.

When important values are at stake and only limited time is available, decision makers often need to act without waiting for more complete information. The high level of health protection set in the EU legislation presupposes that – according

---

[22] *Ibid.*, at 53.
[23] *Ibid.*, at 54.

to well-established case law of the EU courts – the protection of public health takes precedence over economic considerations.[24] Under the General Food Law, however, the lack of complete information in a situation in which consumer health or safety are in danger does not necessarily give grounds for invoking the precautionary principle in risk management. To explain this, we will refer in more detail to the concept of risk management.

The aim of risk management is to choose, on the basis of risk assessment, between different available legislation and control options to minimize or eliminate risks that are too high for society. Control options encompass a broad range of enforcement tools for removing unsafe food from the market and penalizing responsible parties. The planning of the allocation of (limited) inspection and enforcement resources in accordance with risk assessment enhances the effectiveness of risk reduction. In some cases, tightening controls can be an adequate response to reduce the risk to an acceptable level.

Once the appropriate options have been selected from the broad range of possible risk management tools, they have to be implemented. The alert message sent in the *Malagutti* case is part of the implementation of the risk management measure fixing the maximum level for a pesticide residue, dicofol, in fruit.[25] Therefore, it is assumed that the health risks of dicofol had been assessed before the decision concerning the acceptable level of this pesticide residue in fruit was taken and before it was implemented. Thus, the *Malagutti* case did not deal with *scientific* uncertainty, but with a lack of complete information about the factual situation. The distinctive feature of the precautionary principle is not uncertainty as such, but *scientific uncertainty* – a situation when scientific evidence is inconsistent or inconclusive to such an extent that it does not allow the risk to be determined with sufficient certainty. Under this analysis, the alert could not be based on the precautionary principle as codified in Article 7 GFL.

4.5 The role of science and other legitimate factors in the application of the precautionary principle

4.5.1 Risk assessment and other legitimate factors

Food safety policy is inextricably linked to risk assessment, but decision making needs to reflect a wide array of social values. Although the nature of food safety risks makes science a necessary element in the analysis of risk, science alone is not sufficient in the risk management process (Ansell and Vogel, 2006b).

---

[24] See e.g. *Solvay Pharmaceuticals BV* v. *Council, supra* note 17, at 121; and *Pfizer, supra* note 13, at 456.
[25] Dir. 2000/42/EC of 22 June 2000 amending the Annexes to Dir. 86/362/EEC, Dir. 86/363/EEC and Dir. 90/642/EEC on the fixing of maximum levels for pesticide residues in and on cereals, foodstuffs of animal origin and certain products of plant origin, including fruit and vegetables respectively, OJ 2000, L 158/51.

A possible conflict of values often results from differences between scientific risk assessment and the perception and acceptance of risk by society. Whereas risk assessment is a hazard evaluation made by experts, laypersons depend on the perception of risk, which is intuitive and relies on images and associations, connected by experience with a feeling whether something is good or bad (Slovic, 1987, 2000). However, saying that the perception of risk is irrational as opposed to scientific and rational risk assessment is too much of a simplification. Slovic *et al.* (2004) note that risk perception 'enabled human beings to survive during their long period of evolution and remains today the most natural and most common way to respond to risk'. In fact, both systems should complement rather than oppose each other.

The relation between science and politics reflects the interplay between risk assessment and risk perception, but also relates to the role of factors other than natural science in the risk management process and in the risk assessment procedure itself. In fact, the term risk assessment can have narrower and broader meanings. Some identify the term with quantitative risk assessment; others include qualitative expressions of risk. The broadest understanding of risk assessment goes further and embraces analysis and comparison concerning different regulatory strategies, as well as economic and social consequences of these regulatory decisions (NRC, 1983: 18).

The European Commission's Scientific Steering Committee (SSC) dealt with the inclusion of factors other than natural science in risk assessment. In its 2000 First Report on the Harmonization of Risk Assessment Procedures, the Committee concluded that a number of important issues that are currently not taken into account should be incorporated into the formal risk assessment process. The Committee identified three main issues: animal welfare, sustainability, and human quality of life parameters (SSC, 2000: 110).

In the Committee's view, the concept of quality of life is multidimensional. Its starting point was the WHO definition of health as 'a state of complete physical, social and mental well-being, and not merely the absence of disease or infirmity...' (SSC, 2003: 3). In consequence, apart from the classic physical, biological, medicinal or chemical scientific areas, other factors such as cultural, economic, social, psychological and ethical issues should be incorporated into risk assessment. Thus, the risk assessment framework should be multidimensional and include also social scientists in the process.

The General Food Law recognizes that in some cases societal, economic, traditional, ethical and environmental factors and the feasibility of controls should legitimately be taken into consideration in selecting appropriate food safety policy options.[26]

---

[26] Rec. 19 GFL.

These factors, however, apply to risk management, rather than to risk assessment as the SSC suggests.

To what extent can the precautionary principle be based on these factors? Can some quality of life aspects be included in the decision-making process by the application of the precautionary principle (e.g. public risk perception in the case of genetically modified organisms – GMOs)? In the SSC Reports, the authors suggested that sustainability is one of the concerns that are likely to be dealt with by reference to the precautionary principle (SSC, 2000: 118). Under the definition of the principle, this would be possible, however, only under the condition of scientific uncertainty and not where a risk can be fully identified.

The definition of the precautionary principle refers to persisting 'scientific uncertainty' as one of the conditions of the application of the principle. The 'scientific uncertainty' seems to refer to risk assessment, since precautionary measures have to be reviewed as soon as new scientific information 'needed to clarify the scientific uncertainty and to conduct a more comprehensive risk assessment' is available.

The distinction between 'scientific' evaluation, on which the application of the precautionary principle depends, and other legitimate factors, which come to play during the risk management stage, becomes blurred if the formal risk assessment process incorporates animal welfare, sustainability, and human quality of life parameters and includes social science as proposed in the SSC reports. To explain possible theoretical consequences of this inclusion, we will refer to the *Pedicel* case, decided by the EFTA Court in 2005, concerning the interpretation of the free movement of goods and services rules within the European Economic Area (EEA).[27]

4.5.2 Uncertainties over wine advertising

The *Pedicel* case arose in the context of a dispute over the decision imposing a sanction against a company engaged in publishing 'Vinforum', a Norwegian magazine for wine connoisseurs. The December 2003 issue contained commercial wine advertisements, a breach of the prohibition against alcohol advertising under Norwegian law, and the Directorate for Health and Social Affairs in Norway imposed a fine on the publisher. The questions referred to the Court were the following: can such a prohibition be maintained out of concerns for public health, and if so, to what extent would the application of the precautionary principle in this field be in conformity with the case law of the EFTA Court and European Court of Justice?

---

[27] EFTA Court Case E-4/04, *Pedicel AS v. Directorate for Health and Social Affairs*, [2005] EFTA Court Report 1.

The effects of excessive alcohol consumption on human health considered from a toxicological point of view are well known, and – clearly – do not require recourse to the precautionary principle. The question raised before the Court, however, concerned a more specific issue: uncertainty with regard to the assessment of the effects of alcohol *advertising* on the consumption of alcoholic beverages. If scientific evidence concerning such effects on society is not conclusive, would it be possible to prohibit this activity provisionally – pending more comprehensive scientific information – on the basis of the precautionary principle? This leads us to a more general question: to what kind of scientific uncertainty does the precautionary principle refer? Is it limited to certain specific scientific domains?

In the judgment, the EFTA Court merely stated that 'such uncertainty did not arise in a domain which would allow for the invocation of the precautionary principle as developed in the case law of the three courts'.[28] However, the Court did not provide any further explanation in which domains the precautionary principle could be applied.

4.5.3 The scope of risk assessment

To analyse the question referred to the EFTA Court in light of the definitions set out by the General Food Law, we need to focus on the scope of the risk assessment and its impact on the precautionary principle.[29]

Article 6 GFL recognizes that 'food law shall be based on risk analysis except where this is not appropriate to the circumstances or the nature of the measure'. This provision clearly indicates that some areas of Food Safety Regulation remain outside the risk analysis framework. Recital 17 GFL stipulates that risk analysis provides a methodology to determine effective, proportionate and targeted measures in cases where food law is aimed 'at the reduction, elimination, or avoidance of a risk to health'. Therefore, in general, the obligation to base food law on risk analysis is limited to measures aimed at the protection of health.

Clearly, not all measures require recourse to scientific evidence. As an example, the Commission's proposal for the General Food Law[30] mentions law relating to consumer information and the prevention of misleading practices. It does not mean, however, that all measures related to consumer information are automatically excluded from the scope of risk analysis. Although nutrition and health claims

---

[28] *Ibid.*, at 60.

[29] However, it should be noted that this case – as well as other examples discusses in this article – was brought to court before enactment of the GFL. The GFL was incorporated into the EEA Agreement in 2007 by Decision of the EEA Joint Committee No. 134/2007 of 26 Oct. 2007 amending Annex I (Veterinary and phytosanitary matters) and Annex II (Technical regulations, standards, testing and certification) to the EEA Agreement, OJ 2008, L 100/33.

[30] COM(2000) 716 final.

are consumer information, they must be based on risk assessment.[31] Besides, risk reduction measures may include information about adequate treatment or specific recommendation for population at risk on the label, in presentation or advertising, as an alternative that is less restrictive to trade than a total ban (EC, 2000b: 18). Hence, although the prohibition of advertising of alcoholic beverages is a measure related to consumer information, it aims at the reduction of a risk to health and therefore it can still be considered within the risk analysis methodology.

According to the General Food Law, 'risk' is 'a function of the probability of an adverse health effect and the severity of that effect, consequential to a hazard'.[32] 'Risk assessment' consists of four elements: hazard identification, hazard characterization, exposure assessment, and risk characterization.[33] The Regulation defines 'hazard' as 'a biological, chemical or physical agent in, or condition of, food or feed, with the potential to cause an adverse health effect'.[34] These definitions seem to indicate that all elements of risk assessment are related to *biological*, *chemical* or *physical* agents and focus on scientific aspects only. Examples falling under this definition may include toxic substances, radiation, bacteria, viruses, and other pathogens. Consequently, the concept of risk assessment set out in the GFL leaves outside its scope aspects of risks – and uncertainties related to them – that are not related to natural science.[35] It follows that the precautionary principle as codified in Article 7 GFL cannot be applied to scientific uncertainty concerning the effects of advertising of alcohol on society because it is not a type of scientific uncertainty that can be evaluated during the risk assessment process.

In the risk analysis model developed by the Codex Alimentarius Commission, the separation between risk assessment and risk management should be respected. For this reason, other legitimate factors that may be taken into account at the risk

---

[31] Reg. 1924/2006 OJ 2007, L 12/3 (Corrigendum).
[32] Art. 3(9) GFL.
[33] Art. 3(11) GFL.
[34] Art. 3(14) GFL. 'Exposure assessment' is not defined in the GFL, but the risk analysis methodology referred to in the Reg. is based on Codex Alimentarius standards, which define exposure assessment as 'the qualitative and/or quantitative evaluation of the likely intake of biological, chemical, and physical agents via food as well as exposures from other sources if relevant', whereas 'risk characterization' is 'an estimation based on hazard identification, hazard characterization and exposure assessment' (CAC, 2007a). The Risk Analysis Principles Applied by the Codex Committee on Food Additives and the Codex Committee on Contaminants in Foods stipulate that risk assessment outputs is limited to presenting deliberations and conclusions of risk assessment and safety assessment. The communication of risk assessment to risk managers should not include the consequences of the Committees' analyses on trade or other non-public health consequence. *Ibid.*: 123.
[35] EFSA's mandate was especially highlighted in its opinion on animal welfare aspects of the killing and skinning of seals. EFSA explicitly stated that it looked at the 'animal welfare aspects of the methods currently being used for killing and skinning seals', and that, in drafting the scientific opinion, the panel 'did not take into consideration any ethical, socio-economic, human safety, cultural or religious or management issues, the emphasis has been to look at the scientific evidence and to interpret that in the light of the terms of reference' (EFSA, 2007).

management stage should not affect the scientific basis of risk analysis. Because the risk assessment model applied in the General Food Law does not capture social, economic or cultural aspects related to a risk, scientific uncertainty in these domains cannot constitute a basis for application of the precautionary principle. These factors, however, do contribute to the final decision-making process at the risk management stage, and their role seems to be especially important in situations where scientific evidence is inconclusive, insufficient or uncertain.

4.6 The precautionary principle as a directing principle

Different interpretations of the concept of precaution, however, do not depend only on whether it is applied at the risk assessment or risk management stage and do not relate only to different concepts of scientific uncertainty. Above all, the precautionary principle may be expressed in different ways at different policy levels.

In the context of Australian environmental law, Jones and Bronitt distinguish two ways of including precaution in legislation: as an express obligation by the incorporation of precaution in a legal act, or by creating a burden or obligation that in effect creates an implied precautionary policy (Jones and Bronitt, 2006). In the TFEU – as was the case in the EC Treaty – the prevention, precautionary and polluter pays principles are directing principles (de Sadeleer, 2002). They guide public authorities and become operative through specific regulatory and legislative actions, which are subordinated to these principles. The wide application of the precautionary principle explains the lack of a definition in the Treaty. The implementation of the principle results in different legal instruments, depending on the area of regulation (De Sadeleer, 2002: 12). Between a guiding precautionary principle formulated in a general manner and a normative precautionary principle, which applies directly, a series of successive regulatory expressions of the principle may occur, gradually making it more concrete (De Sadeleer, 2002: 310).

We will use a classic example of a regulatory framework based on the precautionary principle – GMOs – to explain these different levels and meanings of precaution and their relation to the precautionary principle outlined in the General Food Law.

4.6.1 Precautionary regulatory framework: pre-market approvals of GM products

Regulation 1829/2003 lays down EU procedures for the authorization and supervision of food or feed containing, consisting of or produced from genetically modified organisms (GMOs).[36] The regulatory framework for GMOs assumes that food or feed containing or consisting of organisms in which the genetic material

---

[36] OJ 2003, L 268/1.

(DNA) has been altered in a way that does not occur naturally may pose greater danger to human health than conventional food/feed. In consequence, GMOs require a stricter regulatory regime and control: the Regulation establishes a case-by-case and step-by-step procedure before GM products are placed on the market. The authorization procedure is centralized and valid throughout the EU. GM food or feed can only enter the European market if it has been documented to be as safe to human health as conventional food/feed.

The Regulation on GM food and feed is just one example of prior approval procedures in the EU food safety regime. Prior approvals (or 'positive lists') exist in European legislation for certain substances such as food additives, pesticides, pharmaceutical products, or foods fortified with vitamins or minerals. This procedure places the burden of producing scientific evidence on food business operators. While – as a general rule – it is up to society (public authorities, consumer organizations, citizens, etc.) to demonstrate that a product or process may be dangerous to human health, the philosophy behind pre-market approvals is to consider a product dangerous until the proponent of placing it on the market proves otherwise. Thus, the burden lies with the applicants to prove safety and not with others to prove harm (Levidow *et al.*, 2005: 268). One of the reasons of the reversal of the burden of proof is that in some cases such scientific evidence may not be available, or at least could be more difficult to obtain for someone from outside. Besides, in some cases, the person who has economic interest in placing the product on the market, and not society, should bear the costs of producing scientific evidence.

Pre-market approvals are called precautionary measures because they are considered to take a normative stand on certain categories of foodstuffs: the foodstuffs are deemed *a priori* dangerous. A substance or process can be authorized only when the risk to human health can be assessed with sufficient scientific certainty. The Commission states in its Communication that, measures based on the precautionary principle may assign responsibility for producing the scientific evidence necessary for a comprehensive risk evaluation, and stipulates that prior approval procedure is 'one way of applying the precautionary principle, by shifting responsibility for producing scientific evidence' (EC, 2000b: 22). Also the European Courts see prior authorization as 'one of the possible ways of giving effect to the precautionary principle'.[37]

4.6.2 Individual authorizations of GM products

The policy framework for GM food and feed, albeit precautionary by its very nature, does not prejudge the safety of individual GM products. Every application under Regulation 1829/2003 must undergo risk assessment carried out by EFSA (European Food Safety Authority). The Commission's proposal for granting or

---

[37] *Pfizer, supra* note 13, at 145.

refusing authorization of a GM product is based on EFSA's evaluation. If the Commission decides to diverge from EFSA's opinion, it must give reasons for it. A single risk management process involves the Commission and the Member States through the comitology procedure. In a situation of scientific uncertainty, the measure can be based on the precautionary principle.

Within the framework of the GM Food and Feed Regulation, authorities often invoke the precautionary principle in emergency measures concerning products that have entered the market without authorization. The 2006 Commission Decision concerning non-authorized genetically modified organism 'LL RICE 601' in rice products[38] is an example of such a measure. Informed by the US authorities about the GM rice found in the food and feed supply in the US and possible contamination of exports to the EU, the Commission asked EFSA to assess the risk related to this GM rice on the basis of a summary risk assessment provided by the US authorities. Meantime, the Commission took an emergency measure requiring imports of US rice to be certified as free of LL RICE 601.

The scientific evaluations concerning the GM rice presented by the US Department of Agriculture and the US Food and Drug Administration,[39] as well as by the advisory bodies in the UK[40] and in the Netherlands,[41] and available at the moment when the Commission was taking the emergency measures, concluded that the rice posed no risk to human health. These findings were confirmed by EFSA stating that – although the available data were not sufficient to assess the safety of the GM rice in accordance with the EFSA guidance for risk assessment – the consumption of imported rice containing trace levels of LL RICE 601 was not likely to pose an imminent safety concern to human health. Yet, the Commission's measure requiring certification of rice imported from the US was upheld,[42] based on the precautionary principle, resulting from the presumption of risk for non-authorized products:

> Since genetically modified rice 'LL RICE 601' is not authorized under Community legislation and in view of the presumption of risk on products not authorized according to Regulation (EC) No. 1829/2003, which takes into account the precautionary principle laid down in Article 7 of Regulation (EC)

---

[38] Dec. 2006/578 of 23 Aug. 2006 on emergency measures regarding the non-authorized genetically modified organism 'LL RICE 601' in rice products, OJ 2006, L 230/8.

[39] USDA (2009).

[40] FSA (2009).

[41] RIVM-RIKILT (2009).

[42] Dec. 2006/601 of 5 Sept. 2006, OJ 2006, L 244/27; Dec. 2006/754 of 6 Nov. 2006, OJ 2006, L 306/17. In 2007, the US Department of Agriculture (USDA) submitted a proposal of protocol to the Commission that would ensure that the products in question are subject to official sampling by US authorities and that the consignments of those products would be accompanied by the original report indicating that LL Rice 601 was not detected. In consequence, mandatory sampling and analysis at the points of entry into the Community was abolished (Dec. 2008/162 of 26 Feb. 2008, OJ 2008, L 52/25).

No. 178/2002, it is appropriate to take emergency measures to prevent the placing on the market in the Community of the contaminated products.[43]

4.6.3 Is the GM Regulation based on the precautionary principle?

Although legal doctrine and policy documents, as well as the EC Communication, seem to consider reversal of the burden of proof as a core component of the precautionary principle, the definition of the principle in the General Food Law does not mention a shifting of a burden of proof as a consequence of its application. Nor does the application of the precautionary principle change anything in the general principles in the matter of proof.

The concept of pre-market approvals is built on the assumption that a product is unsafe until proven otherwise. However, to say that it is based on the precautionary principle is to apply the principle beyond the context of the risk analysis framework established in the General Food Law. The precautionary principle is an instrument that allows decision makers to take measures in the condition of scientific uncertainty or, in the Court's words, 'without having to wait until the existence and gravity of those risks become fully apparent'.[44] The General Food Law defines the precautionary principle and sets out three pre-requisites for its application.[45] Similarly, even before the General Food Law entered into force, the European judiciary had held that the precautionary principle applies to situations where risk assessment does not provide decision makers with 'conclusive scientific evidence of the reality of the risk and the seriousness of the potential effects were that risk to become a reality',[46] and that 'a preventive measure cannot properly be based on a purely hypothetical approach to the risk, founded on mere conjecture which has not been scientifically verified'.[47]

Within the GM Food and Feed Regulation, the precautionary principle as defined in the GFL may thus be invoked to justify decisions on individual authorizations, where risk assessment remains inconclusive, but the possibility of harmful effects must be 'identified, following assessment of available information', and not just hypothetical. The example of the GM rice shows that the precautionary principle may be triggered by uncertainty resulting from the lack of some relevant scientific data. This lack can be linked to urgency and emergency situations such as inadvertent release of a product that has not been authorized on the market. Even in this type of situation, however, recourse to the precautionary principle

---

[43] Rec. 5 Dec. 2006/578.

[44] Case C-41/02, *Commission* v. *The Netherlands*, [2004] ECR I-11375, at 52.

[45] See Section 4.3.2.

[46] *Pfizer, supra* note 13, at 142.

[47] *Ibid.*, at 143.

still presupposes *identification* of possible harmful effects on health, following assessment of available information, and not just a hypothetical presumption.[48]

Another scenario resulting from the pre-market approval procedure is a situation where the proponent of an activity provides enough scientific information to conduct a comprehensive risk assessment. Public authorities have sufficient information available to take a decision and do not have to resort to the precautionary principle at all. In this case, no scientific uncertainty persists, except that responsibility for providing the scientific evidence necessary for a full risk assessment is on the producer. Is shifting the burden of producing scientific evidence enough to call it scientific uncertainty? Why should such a procedure be referred to as based on the precautionary principle then?

4.7 Conclusion

Although the European approach placing the precautionary principle at the risk management stage leaves a broader margin of discretion to decision makers than the 'precautionary approach' included in risk assessment, this discretion has some clear limitations determined by the risk analysis framework set out in Regulation 178/2002.

Environmental law continues to shape the precautionary principle applied in the field of food safety. However, the principle has different legal status in the two fields. Environmental measures are based on the precautionary principle formulated in a general manner in the TFEU, and previously in the EC Treaty. This principle becomes normative through secondary legislation in different ways. Its expression will be different in the field of waste water treatment, the use of hazardous materials or the protection of endangered species. On the contrary, the General Food Law establishes a comprehensive EU food safety policy based on the general principles applying to virtually all food safety legislation at EU and national level. The precautionary principle is one of these principles and part of risk the analysis methodology for determining efficient measures based on scientific evidence. Being a directing principle for public authorities, at the same time, however, it is concrete enough to exclude from its scope some food safety regulatory frameworks, which – although undoubtedly having 'protective' or even 'precautionary' character – do not meet the criteria set out in the definition. This creates terminological confusion and leads to a paradox: the GM Regulatory framework – one of the most celebrated examples of the application

---

[48] See also *Pfizer*, cited *supra* note 13, at 144: 'It follows from the Community Courts' interpretation of the precautionary principle that a preventive measure may be taken only if the risk, although the reality and extent thereof have not been "fully" demonstrated by conclusive scientific evidence, appears nevertheless to be adequately backed up by the scientific data available at the time when the measure was taken'; and at 155: 'A scientific risk assessment must be carried out before any preventive measures are taken'.

of the precautionary principle, placed at the intersection of environmental and food safety regulations – risks non-compliance with the precautionary principle defined in the EU food safety policy. Although all the examples discussed in this article contain an implied precautionary policy, precaution as a general notion associated with a high level of health protection is not always equivalent to the precautionary principle defined in the General Food Law, and for these other issues the term 'precautionary principle' seems to be no longer the same.

5. Different actors, different factors?

Science and other legitimate factors in EU and national food safety regulation

Abstract

According to the principle of risk analysis established by Regulation 178/2002, food safety measures in the EU and Member States must be based on scientific risk assessment. Apart from science, however, decision makers should take into account other legitimate factors, such as societal, ethical or traditional concerns. The extent to which risk managers can deviate from scientific evaluations in considering these factors depends on how much discretion is conferred on public authorities. This chapter compares the discretion at both national and Union levels of food safety regulation in the context of the internal market mechanism by analysing the standard review applied to food safety measures by the European judiciary.

5.1 Introduction

Risk regulation interferes with market mechanisms to protect interests such as health and the environment (Hood *et al.*, 2001).[1] Regulation in these fields relies heavily on technical expertise. Risk analysis is a methodology incorporating science into the decision making process, where risk managers take regulatory decisions based on the scientific risk assessment provided by experts.

Clearly, science is not able to take responsibility for all decisions for society. Apart from scientific risk assessment, decision makers must also take into account different interests and concerns of social, economic or ethical character, including, e.g. consumer risk perception, animal welfare, and traditional methods of production. In the risk analysis methodology established by Regulation 178/2002 laying down the general principles and requirements of food law, establishing

* Reprinted from European Journal of Risk Regulation, 4, Szajkowska, A., Different Actors, Different Factors - Science and Other Legitimate Factors in EU and National Food Food Safety Regulation, 523-539, 2011, with permission from Lexxion.

A modified version of this chapter has been published as:
Szajkowska, A., 2010. Food safety governance from a European perspective: risk assessment and non-scientific factors in EU multi-level regulation. In: Hospes, O. and Hadiprayitno, I. (eds.) Governing food security. Law, politics and the right to food, Wageningen Academic Publishers, Wageningen, the Netherlands, pp. 201-229.

---

[1] Some authors refer to 'social regulation' to describe this process – see Joerges (1997).

the European Food Safety Authority and laying down procedures in matters of food safety (the so-called General Food Law, hereinafter also referred to as the GFL)[2] these concerns, considered at the risk management stage, are recognized as 'other legitimate factors'. Despite the growing controversy about whether and to what extent risk managers, by taking into account non-scientific factors, may decide against the findings of risk assessment, the role of these factors in EU food safety governance is not clearly formulated.

The extent to which risk managers can deviate from science in highly technical areas of risk regulation is in fact the question of how much discretion is conferred on public authorities. The risk analysis methodology set out in the General Food Law does not operate in a regulatory vacuum. EU and national food safety regulation must be interpreted in light of the rules concerning the free movement of goods.[3] Hence, ultimately, the manner in which such discretion is exercised comports with the Treaty framework relating to the functioning of the internal market and the standard review applied by the European judiciary.

This chapter examines how the internal market mechanism and Treaty provisions relating to the free movement of goods determine the way other legitimate factors are incorporated into the decision making process in EU multi-level food safety regulation. It starts by describing shortly the division of competences between the EU and the Member States in this area and standard-setting mechanisms in the context of the internal market. Then, an overview of the main elements of the food safety scientific governance established by the General Food Law will be given. Next, we discuss the principle of risk analysis in light of the standard of judicial review of EU measures aimed at the protection of human health. Subsequently, we analyse the discretion applied to national measures, with a special focus on the relation between EU and national scientific opinions. Lastly, we look at the role of other legitimate factors, referring to two possibilities resulting from their inclusion in the decision-making process: standards ensuring a level of protection that is lower than that recommended by risk assessors, and a level of protection higher than that advocated by scientific opinion. Although the focus of this chapter is the risk analysis model established in the General Food Law, we also refer to examples from other areas of risk regulation (legislation for hazardous substances and environmental standards). The chapter shows that, although the risk analysis methodology established by the General Food Law applies to all food safety legislation, the role of science and other legitimate factors determined by the context of the internal market and the scope of judicial review is different for EU and national measures.

---

[2] OJ 2002, L 31/1.

[3] See Case C-47/90, *Delhaize* [1992] ECR I-3669, para. 26; Case C-315/92, *Clinique* [1994] ECR-317, para. 12.

5.2 Division of powers between the EU and the Member States

5.2.1 Treaty framework for EU food safety regulation

In EU multi-level food safety regulation, the division of competences between the Member States and the Union is shaped mainly by the Treaty provisions relating to the functioning of the internal market. In principle, the free movement of goods is ensured through the principle of mutual recognition, developed in the Cassis de Dijon case law.[4] This principle breaks up national standards, precluding a Member State from prohibiting the sale of a product which does not conform to domestic regulation, but which has been lawfully marketed in another Member State. As a consequence, it is no longer necessary to harmonise all standards and compositional rules for all food products to ensure their free circulation in the internal market.

However, the Treaty on the Functioning of the European Union (TFEU) in Article 36 (ex Art. 30 EC) provides exceptions to the free movement of goods, i.e. on grounds of the protection of health and life of humans, animal or plants.[5] Thus, in the area of food safety, where the main purpose of regulation is the protection of human life and health, Member States can invoke Article 36 TFEU as a basis for their legislation and exemption from mutual recognition. As a consequence, harmonisation of national food safety laws still remains the only way to assure the free movement of foodstuffs – national competences are pre-empted through Article 114 TFEU (ex Art. 95 EC). Nearly all legislation related to food safety is now harmonised at the EU level, whereas technical food standards remain mostly within the competence of the Member States.

5.2.2 Multi-level regulation through Article 114 TFEU

Because food safety regulation is to a large extent harmonised at EU level, Article 114 TFEU in fact delimits competences between the EU and Member States in this field.[6] The Article envisages three possibilities for national measures to derogate from harmonised European regulation.

Paragraph 4 allows Member States to *maintain* more stringent national provisions existing prior to harmonisation of laws at the EU level on grounds of major

---

[4] Case 120/78, *Rewe-Zentral AG* v. *Bundesmonopolverwaltung für Branntwein* [1979] ECR 649.

[5] The other exceptions are: public morality, public policy or public security; the protection of national treasures possessing artistic, historic or archaeological value; and the protection of industrial and commercial property.

[6] On Art. 114 TFEU see De Sadeleer (2003), Dougan (2000), Hanf (2001) and Vos (2001). In 2002 the Commission issued a Communication concerning Article 95 EC (EC, 2002b).

needs listed in Article 36 TFEU, or referring to the environment or the working environment (see Annex 1).

Paragraph 5 provides the possibility to introduce *new* national measures after harmonisation. This possibility, however, is limited to the environment or the working environment – paragraph 5 does not refer to Article 36 TFEU, and therefore the introduction of new national measures derogating from harmonised EU legislation in the area of food safety is not possible. The new national measures must be based on new scientific evidence relating to the protection of the environment or working environment on grounds of a problem specific to that Member State and arising after the adoption of the harmonisation measure.

National provisions derogating from EU harmonisation measures are subject to the Commission's approval (Art. 114(6) TFEU). They must be necessary and proportionate to the objective pursued, cannot constitute a means of arbitrary discrimination, a disguised restriction on trade or an obstacle to the functioning of the internal market.[7]

Paragraph 10 introduces the possibility to include safeguard clauses in harmonisation measures. The safeguard clauses permit Member States to restrict or suspend trade in foodstuffs for one or more non-economic reasons referred to in Article 36 TFEU. The national safeguard measure must be temporary and subject to a Union control procedure. However – unlike the other two possibilities discussed above – this provision is addressed to the EU legislature and does not confer on the Member States any powers *per se*.

In addition, apart from safeguard clauses, harmonisation may allow other derogations or leave some aspects of the subject matter to the Member States. This possibility will be discussed at the end of this chapter.

The opt-out clauses introduced in paragraphs 4 and 5 are similar to the pattern of minimum harmonisation, setting a threshold that standards have to meet, but giving Member States the possibility to retain (paragraph 4) or to establish (paragraph 5) more stringent measures. By introducing these clauses the Single European Act took a more flexible approach to the creation of the internal market, and actually, in the beginning, the new harmonization provisions, especially paragraph 4, raised fears of jeopardizing free movement of goods (Pescatore, 1987). Consider the conditions set out in paragraphs 4 and 5. The requirement that national measures be based on 'new scientific evidence on grounds relating to a problem specific to that Member State' does not exist in paragraph 4. In his opinion in the case *Denmark* v. *Commission*, Advocate General Tizzano stressed that paragraph 4 needed to be interpreted strictly, requesting justifications consisting

---

[7] See, e.g. Dec. 1999/5/EC, OJ 1999, L 3/13, para. 15; Dec. 1999/830/EC, OJ 1999, L 329/1, para. 18.

in a situation specific to the Member State concerned, otherwise it would open the possibility to maintain or introduce a derogation from the principle of unity of the market, on the basis of a 'unilateral assessment of the need for raising the standard'.[8] He went on to conclude that:

> [T]o allow such a claim would mean transforming Article 95(4) EC into a veritable permanent opt-out clause from any harmonisation directive, in stark contrast with the principles and purposes of the system and with the logic which, in the protection of the general interest, inspires the division of powers between the Community and the Member States.[9]

The Court did not confirm these arguments, ruling that 'neither the wording of Article 95(4) EC nor the broad logic of that article as a whole entails a requirement that the applicant Member State prove that maintaining the national provisions which it notifies to the Commission is justified by a problem specific to that Member State'.[10]

In practice, however, few Member States have used Article 114(4) TFEU to maintain their national provisions.[11] The limited number of national notifications may be explained by the fact that the very threat of invoking the opt-out clause compels the Member States to seek a harmonisation measure that ensures a level of health protection satisfactory for all States (Hanf, 2001). In fact, Article 114(3) TFEU explicitly stipulates that in the field of health, safety and consumer protection the Union institutions are obliged to take as a base a high level of protection, taking into account any new development based on scientific facts. The relationship between national and EU standards based on Article 114 TFEU will be discussed in more detail in the next section.

5.2.3 Race to the top[12]

In Grey's words, in the EU harmonization system defined by Article 114 TFEU, '[e]ach national act calls for a Community act to resolve the situation' (Grey, 1990). In many cases, the initially differentiated standards lead to a uniform regulation: the internal market mechanism eventually brings Union legislation in line with national standards. Consequently, national derogations help develop stricter Union environmental and food safety standards and lead to increased involvement in the field of the environment, food safety, and consumer protection. The European Economic and Social Committee (EESC) in its opinion on the relation between the articles referring to the internal market and environmental protection stated:

---

[8] Opinion of A.G. Tizzano of 30 May 2002 in Case 3/00.
[9] *Ibid.*, para. 76.
[10] Case 3/00, *Denmark* v. *Commission* [2003] ECR I-2643, para. 59.
[11] For an overview of national derogations see Onida (2006).
[12] The title refers to Otsuki *et al.* (2001).

> The practical implementation of these articles involves both approaches at the same time, i.e. harmonization and 'non-harmonization'; by not harmonizing and opting for a higher level of protection, Member States force a constant upwards adjustment of environmental standards. This leads to a halfway house situation where there are technical barriers to trade, but these barriers may help to reach the ultimate aim – environmental protection (EESC, 1998).

This phenomenon was described by David Vogel in 1995 in relation to the history of United States automobile emission standards. In 1970, the Clean Act Amendments expressly allowed California to introduce stricter standards than those required for the rest of the US. In 1990, the US emission standards became stricter following California's standards. Congress once again permitted California to set stricter standards, and gave the other states the possibility to choose either national or California standards. A few years later 12 states asked the federal government permission to introduce California's standards. The new standards became the basis for setting new minimum federal requirements. Thus, instead of a race to the bottom to enhance competitiveness and attract investment, standards in the internal market are eventually becoming a uniform regime in line with the strictest requirement (Vogel, 1995).

The 'California effect' can be observed in the EU regulation on creosote, a wood preserving agent.[13] Directive 94/60/EC amending Directive 76/769/EEC on the marketing and use of certain dangerous substances harmonised the use of creosote and laid down restrictions on its marketing.[14] Despite the EU restrictions, four Member States – the Netherlands, Germany, Sweden, and Denmark – decided to maintain their stricter national provisions (on the basis of Article 114(4) TFEU). In this case, following the national requests for derogations, the Commission ordered additional studies and asked the Scientific Committee on Toxicity, Ecotoxicity, and the Environment (SCTEE) to evaluate the findings. The Committee concluded that there was a cancer risk to humans from creosote containing benzo[a]pyrene at the concentration set out in the Directive, although the magnitude of the risk could not be estimated with certainty. In consequence, the Commission approved all four national derogations in light of the precautionary principle.[15]

---

[13] Creosote is principally used for railway sleepers, poles for electricity transport, hydraulic constructions or fences. It is highly toxic and its release to the environment can cause harm to animals and wildlife. Exposure to creosote is also dangerous to human health as it can be carcinogenic.

[14] OJ 1994 L 365/1. Creosote was not permitted for wood treatment if it contained benzo[a]pyrene above a concentration of 0.005% by mass and water extractable phenols at a concentration greater than 3% by mass. By way of derogation, in industrial installations creosote containing benzo[a]pyrene of a concentration of less than 0.05% was allowed.

[15] Dec. 1999/832/EC, OJ 1999, L 329/25; Dec. 1999/833/EC, OJ 1999, L 329/43; Dec. 1999/834/EC, OJ 1999, L 329/63; Dec. 1999/835/EC, OJ 1999, L 329/82.

The review of the studies on health effects of creosote and the approval of the national measures resulted in a re-consideration of the EU harmonisation measure in line with the national derogations. In 2001, the Commission amended Directive 76/769/EEC.[16] The marketing and use of creosote was prohibited, except for industrial applications, where the levels previously applied to all wood treatment were allowed.[17]

In 2002, the Netherlands notified the Commission once again about its stricter national provisions on the use of creosote, derogating from the 2001 Directive in that the ban on creosote included industrial applications. The SCTEE confirmed that the environmental risks highlighted in the scientific studies submitted by the Netherlands, relating to the extensive use of creosote-treated wood for riverbank protection, were justified. The Commission approved the measure and once again announced its intention of examining the harmonisation measure in light of the scientific evidence provided by the Netherlands, in accordance with Article 114(7) TFEU.[18]

To evaluate whether national derogations are necessary and proportionate, and whether they do not constitute a means of arbitrary discrimination or a disguised restriction on trade, the Commission refers to scientific assessment. In the creosote case, an external expert was asked to carry out an additional environmental impact study. The Commission, however, emphasized that – according to the Treaty provisions – the responsibility of producing scientific evidence lies with the requesting Member State and that seeking additional justification for the maintenance of the national measures by the Commission cannot constitute a precedent for the future.[19] In most cases, the Commission confines itself to asking EU scientific institutions to give an opinion on the scientific evidence submitted by the national authorities.

This leads us to the core issue in the application of national derogations and the functioning of the internal market in general, *viz.* possible conflicts between national and Union scientific opinions. The Commission, in principle, relies on the advice of its own scientific committees or EU agencies, whereas Member States invoke studies produced by national scientific institutions. These tensions are especially visible in the field of food safety, where the serious mishandling of the BSE crisis in the nineties shattered public trust in the EU scientific advice and food

---

[16] Dir. 2001/90/EC, OJ 2001, L 283/41.
[17] I.e. creosote containing benzo[a]pyrene at a concentration of less than 0.005% by mass and water extractable phenols at a concentration of less than 3% by mass.
[18] Dec. 2002/884/EC, OJ 2002, L 308/30, para. 80.
[19] See, e.g. Dec. 1999/833/EC, *supra* note 15, paras 32-34.

safety system.[20] A series of actions aimed at improving scientific governance of food safety and regaining consumers' trust resulted in the creation of a system of principles and requirements of food law, new procedures in matters of food safety and a new scientific institution – the European Food Safety Authority (EFSA). The next part of this chapter will briefly discuss the main elements of this scientific food safety governance established by Regulation 178/2002.[21]

5.3 Scientific food safety governance

5.3.1 European Food Safety Authority

The European Food Safety Authority (EFSA) was established by the General Food Law as an important element of the new EU policy to improve food safety and restore confidence in the food supply in the EU (European Commission, 2000a). EFSA provides communication on food risks and independent scientific advice to support the Commission, European Parliament and Member States in taking risk management decisions. Accordingly, Scientific Committees attached to the Commission and delivering scientific advice in the field of food safety were reorganised and their role was taken over by EFSA.[22]

EFSA carries out risk assessment underpinning food safety legislation (in particular in so far as it addresses specific foods or substances) in accordance with the principles of excellence, independence and transparency. In the EU legislation introduced after the General Food Law, risk assessment is centralized and entrusted to EFSA.[23] Although EFSA does not have any regulatory powers, its scientific opinions are crucial for drawing up food safety measures within the framework of the risk analysis methodology. Where the draft measure is not in accordance with EFSA's opinion, the Commission must give reasons for its decision.[24] Thus, in exercising its administrative discretion, the Commission has to give careful consideration to the scientific opinion of the Authority. Some authors suggest that this provision may make it difficult for the Commission to diverge from EFSA's conclusions (Alemanno, 2007: 194), but – as the analysis of several examples later in this chapter shows – this seems to be more relevant to national derogations.

---

[20] A special Committee of Inquiry set up by the European Parliament severely criticized the Commission for protecting the market and putting business interests over public health concerns (European Parliament, 1997). On the BSE crisis see Vos (2000a) and Krapohl (2005).

[21] *Supra* note 2.

[22] These Committees were: Scientific Committee on Food, Scientific Committee on Animal Nutrition, Scientific Veterinary Committee, Scientific Committee on Pesticides, Scientific Committee on Plants, and Scientific Steering Committee.

[23] See, e.g. Reg. 1829/2003 on genetically modified food and feed, OJ 2003, L 268/1; Reg. 1924/2006 on nutrition and health claims made on foods, OJ 2007, L 12/3; Reg. 1331/2008 establishing a common authorization procedure for food additives, food enzymes and food flavourings, OJ 2008, L 354/1.

[24] See, e.g. Art. 7(3) Reg. 1331/2008, *supra* note 23.

One of the basic principles of EFSA's mission is to network with national authorities carrying out similar tasks and to facilitate the sharing of knowledge.[25] EFSA is not empowered with a decisive voice in the case of conflict between national and EU scientific opinions. Instead, it is obliged to be vigilant and to identify at an early stage any diverging scientific opinions between the Authority and Member States, and to cooperate with a view to either resolving the divergence or preparing a joint document explaining the differences.[26] Thus, in the network of scientific excellence, EFSA was created to become an authoritative point of reference, whose reputation in scientific matters would 'put an end to competition in such matters among national authorities in the Member States' (Byrne, 2002).

Another procedure designed to deal with diverging scientific opinions is mediation introduced in Article 60 GFL. According to this provision, where a Member State is of the opinion that a measure taken by another Member State is either incompatible with the General Food Law or likely to affect the functioning of the internal market, the Commission may request EFSA to give an opinion on the contentious scientific issue. This procedure is aimed at reaching a 'mutual scientific opinion' and resolving scientific conflicts before the matter is brought before the Court of Justice (see Chapter 2).

5.3.2 Risk analysis methodology: science and other legitimate factors

The General Food Law establishes general principles and requirements of food law that apply to all stages of production, processing and distribution of food and feed.[27] 'Food law' is defined as 'laws, regulations and administrative provisions, governing food in general, and food safety in particular, whether at Community or national level'.[28] Therefore, the principles established in the GFL apply to all legislation in this field at all levels of European governance.

One of the general principles governing EU food safety regulation is the principle of risk analysis, set out in Article 6 GFL. Risk analysis is a widely recognized methodology used for developing food safety standards. The risk analysis approach is used, i.e. in the Codex Alimentarius Commission (CAC) – an inter-governmental body established by the Food and Agriculture Organization (FAO) and the World Health Organization (WHO) to develop international food standards to ensure fair practices in trade and consumer protection in relation to the global trade in food.

---

[25] Art. 36(3) GFL and Reg. 2230/2004 laying down detailed rules with regard to the network of organizations operating in the fields within EFSA's mission, OJ 2004, L 379/64.
[26] Art. 30 GFL.
[27] Art. 4(1) GFL.
[28] Art. 3(1) GFL.

The risk analysis principle requires that measures adopted by the EU and the Member States should generally be science-based.[29] Risk analysis for regulatory actions can be described as a two-step process.[30] During the risk assessment stage, scientists evaluate risks related to a product or technology.[31] The next step is risk management – a political process where decision makers utilize risk assessment findings to develop policy alternatives and select the most appropriate measures to reduce or eliminate risks.[32]

Risk assessment provides an essential input to risk management. Science alone, however, cannot constitute a basis for food safety decisions. Establishing policy in a democratic manner requires taking into account a broad array of societal values and preferences. The creation of public policy in the field of food safety has to reconcile the European patchwork of diverse cultures and traditions concerning food with the single market,[33] which has always been a particularly contentious issue (Ansell and Vogel, 2006b).

Typically such conflicts in the democratic technological society have been described as the science-democracy dichotomy. The politics of food safety creates a paradox of reliance on scientific expertise considered as a device legitimating decision making, which at the same time leads to the distancing of society from political participation in risk governance by failing to respond to other social concerns. Thus, efforts at 'democratizing expertise', where scientific opinions are supplemented by social, ethical or economic concerns expressed by civil society, also have to be interpreted in a broader context of achieving legitimate risk governance by combining science and public participation in the decision making process (Christoforou, 2003; De Marchi, 2003; EC, 2001b; Grundmann and Stehr, 2003; Liberatore and Funtowicz, 2003; Millstone, 2007).

This conflict between scientific considerations and other values translates into the question of how much discretion decision makers enjoy in taking measures

---

[29] The risk analysis methodology is limited to measures aimed at the protection of human health (food safety law). Technical food standards not having as their objective the protection of human health are excluded from the scope of application of risk analysis (see also Chapter 4). On risk analysis see NRC (1983), American Chemical Society (1998), FAO (1997).

[30] The definition of risk analysis, strictly speaking, consists of three components: risk assessment, risk management and risk communication. Risk communication is the interactive exchange of information concerning risks throughout the risk analysis process among risk assessors, risk managers, but also consumers, food businesses and other stakeholders (Art. 3(13) GFL). This element of risk analysis, however, is outside the scope of this study.

[31] Risk assessment comprises four steps: hazard identification, hazard characterisation, exposure assessment and risk characterization (Art. 3(11) GFL). On the application of risk assessment to food safety see McKone (1996).

[32] Art. 3(12) GFL.

[33] As an example, Vogel describes the reaction to the Ruling of the Court of Justice concerning the German 'Reinheintsgebot' (Case 178/84, *Commission* v. *Germany* [1987] ECR 1227). A petition to maintain the national purity decree was signed by 2.5 million German citizens (Vogel, 1995).

in highly technical areas of risk management. The General Food Law recognizes that other legitimate factors, such as societal, economic, traditional, ethical and environmental concerns, and the feasibility of controls should be taken into account when drawing up food safety legislation.[34] The concept of 'other legitimate factors' is also included in the definition of risk management[35] and in Article 6 GFL, which stipulates that, in order to achieve the general objectives of food law established in Article 5 GFL, risk management should take into account the results of risk assessment, other factors legitimate to the matter under consideration and the precautionary principle.

The Regulation merely states that other legitimate factors come into play during the risk management stage. The General Food Law does not provide, however, any indication of how they should be incorporated in the decision making process and to what extent measures based on 'other legitimate factors' can deviate from the conclusions of scientific risk assessment.

5.4 Standard of judicial review of EU and national health protection measures

5.4.1 Proportionality test

The discretion that risk managers enjoy is coupled with the standard of review applied by courts. The rigour or intensity of the scrutiny applied by the judiciary in reviewing acts based on complex technical assessments is a particularly sensitive area. On the one hand, courts do not have a basis for 'second-guessing administrators on such highly technical issues' (Jasanoff, 1994). On the other hand, although the judiciary does not express the desire to carry out scientific assessments on its own, the exercise of administrative discretion in technical areas cannot be totally excluded from judicial review.

In the EU context, the proportionality test, a tool of judicial oversight used to balance some legally recognized or protected interest or right being restricted, typically consists of three elements: a rationality test, necessity test, and proportionality *stricto sensu* (De Búrca, 2000; Emiliou, 1996; Jans, 2000). The rationality test requires that the measure is an adequate means to achieve a purported end, in other words – whether the measure is suitable to protect the interest that it is intended to protect. At this stage Courts can determine whether, e.g. a measure that claims to protect public health is not essentially a disguised restriction on trade. The necessity test goes beyond the simple rationality test and consists in checking whether there exists a less trade restrictive means to achieve the same

---

[34] Rec. 19 GFL.
[35] Art. Art. 3(12) GFL.

objective.[36] The necessity test also applies to the situation where scientific risk assessment shows that a product that has been banned is in fact safe (Jans, 2000). Finally, proportionality *stricto sensu* juxtaposes the costs of introducing a measure and the benefits, and analyses whether the means are proportionate to the objective of the measure.[37] In practice, however, the division is not consistent, and the degree and intensity of review by Courts may differ.

5.4.2 Proportionality of EU food safety measures

In general, the Court of Justice grants the EU institutions a broad discretion in drawing up legislation based on complex technical assessments.[38] The legality of a measure can be affected if the authorities have not complied with the procedural requirements, if the facts established by the authorities are not correct, or if there has been a manifest error of appraisal of those facts or a misuse of powers.[39]

Where the EU institutions enjoy a wide discretion, the administrative procedure becomes an important element of the judicial review. In this regard, the court must establish whether the evidence is accurate, reliable and consistent, and whether it contains all the information necessary to draw conclusions in such complex situations.[40] In the *Pfizer* case, however, the Court of First Instance confirmed that, in cases involving complex assessments, the discretion also applies to a certain extent to the establishment of the facts:

> The Community judicature is not entitled to substitute its assessment of the facts for that of the Community institutions, on which the Treaty confers sole responsibility for that duty. [41]

This is particularly important for EU measures based on risk assessment from EU scientific bodies because, as we will see below, it makes it very difficult to overturn the scientific opinion on which the EU measure is based. In the area of risk regulation the Courts take a relatively modest approach, in general restricted to analysing whether an EU measure is *manifestly inappropriate* to achieve the desired objective.

---

[36] See Case 104/75, *De Peijper* [1976] ECR 613, paras 16-17.

[37] See, e.g. Case 302/86, *Commission* v. *Denmark* [1988] ECR 4607 (*Danish bottles*). In this case, the ECJ found that 'the system for returning non-approved containers is capable of protecting the environment and, as far as imports are concerned, affects only limited quantities of beverages compared with the quantity of beverages consumed in Denmark ... In those circumstances, a restriction of the quantity of products which may be marketed by importers is disproportionate to the objective pursued' (para. 21).

[38] Case T-326/07, *Cheminova* v. *Commission* [2009] ECR II-02685, para. 106.

[39] *Ibid.*, para. 107. See also Joined Cases T-74/00, T-76/00, T-83/00, T-84/00, T-85/00, T-132/00, T-137/00 and T-141/00, *Artegodan and Others* v. *Commission* [2002] ECR II-4945, para. 200.

[40] Case C-405/07 P, *Netherlands* v. *Commission* [2008] ECR I-8301, para. 55.

[41] Case T-13/99, *Pfizer Animal Health SA* v. *Council* [2002] ECR II-3305, para. 169.

Moreover, confronted with divergent scientific opinions of different institutions, the Court has expressed on several occasions confidence in international scientific research, and in particular the findings of the Scientific Committee of Food and later of EFSA.[42] In the *Cheminova* case, for example, the Court had to assess whether the results of a study, suggested by EFSA and undertaken by the company, which established no genotoxicity potential for a substance, could exclude the uncertainties pointed out in the EFSA report. Although the new study addressed the uncertainties, the Court concluded that it could not affect the legality of the decision upholding the ban on that substance because, according to the EFSA report, 'further genotoxicity *studies* (emphasis added) [had] to be provided'. In the Court's reasoning, the mere use of the plural in EFSA's opinion forejudged.[43]

In accordance with the three elements of the proportionality test discussed above, Trachtman argues that proportionality *stricto sensu* is in fact a cost-benefit analysis with a certain margin of appreciation. The margin of appreciation gives more room for all considerations that have to be taken into account during the decision making process, as opposed to merely asking whether the costs outweigh the benefits (Trachtman, 1998). Using the risk analysis terminology set out in the General Food Law, the cost-benefit analysis can be defined as the process of weighing different policy alternatives, taking into account science and other legitimate factors that influence the content of risk management measures.

In the *Pfizer* judgment the Court recognized that 'a cost-benefit analysis is a particular expression of the principle of proportionality in cases involving risk management' and referred to the three elements of the proportionality test:

> [T]he principle of proportionality ... requires that measures adopted by Community institutions should not exceed the limits of what is appropriate and necessary in order to attain the legitimate objectives pursued by the legislation in question, and where there is a choice between several appropriate measures, recourse must be had to the least onerous, and the disadvantages caused must be not be disproportionate to the aims pursued.[44]

The applicant challenged an EU ban on the use of an antibiotic – virginiamycin – as a growth promoter in animals. The rationale for the ban was the protection of human health linked to the probability of transfer of antibiotic resistance developed in animals to humans. Pfizer raised various arguments relating to the benefits of the antibiotic, including animal welfare (whether the use of the antibiotic prevents certain diseases and reduces the mortality rate in animals),[45] economic factors

---

[42] See, e.g. Case 178/84, *supra* note 33, para. 44. On EFSA's de facto superiority over national scientific advice, see also Alemanno (2008).

[43] Case T-326/07, *supra* note 38, para. 146.

[44] Case T-13/99, *supra* note 41, para. 411.

[45] *Ibid.*, paras 420-429.

related to the use of growth hormones in animal farming (possible increased costs to society of such changes in farming methods), and effects of this agricultural practice on the environment.[46]

Although the Court devoted considerable space to these factors, the review remained rather marginal. Despite giving the impression of applying the *stricto sensu* test, the Court confined itself to checking whether the measure was tainted by manifest error, justifying the wide discretion concerning measures protecting human health with the political responsibility. This responsibility concerns policy choices determining the level of protection deemed acceptable for society.[47] In addition, according to settled case-law, public health must take precedence over any other consideration.[48] Hence, it is unlikely that, in assessing risk management measures involving the weighing of different political and economic factors, and protecting a broad public interest, the courts will go beyond the rationality review – the first element of the proportionality test.

Although when assessing the proportionality of EU measures protecting human health Courts do not seem to make a distinction between basic acts (e.g. the directive on the marketing and use of certain dangerous substances laying down restrictions on the use of creosote) and implementing measures (the decision withdrawing the authorisation of virginiamycin), the picture becomes more complicated for national measures derogating from EU legislation, which may require more intensive scrutiny. The next section will discuss the standard of review applied to national derogations in comparison with the discretion enjoyed by the EU institutions, in the light of the risk analysis methodology.

5.4.3 Proportionality of national food safety measures

The Treaty framework for national food safety measures is set out in Articles 34-36 and 114 TFEU. Because Article 36 contains an exception to the rule of the free movement of goods, national authorities have to show that their measures are necessary to protect public health, and that in the view of the results of international scientific research, the product or process in question is dangerous.[49]

In the *Commission* v. *France* judgment, concerning national measures requiring a prior authorization scheme for processing aids and foodstuffs whose preparation involved the use of processing aids, the Court explicitly referred to the risk analysis framework established by Regulation 178/2002 in the context of evidence that national authorities have to provide to justify their measures. Processing aids,

---

[46] *Ibid.*, paras 464-474.
[47] See, e.g. Case 157/96, *National Farmers' Union* [1998] ECR I-2211, para. 61.
[48] Case T-13/99, *supra* note 41, para. 471 and case law referred to therein.
[49] Case C-333/08, *Commission* v. *France* [2010] ECR I-757, para. 87 and case-law referred to therein.

except for extraction solvents,[50] are not harmonized at EU level, hence the French legislation was considered under Articles 34 and 36 TFEU. In this regard, the Court held that:

> A Member State cannot justify a systematic and untargeted prior authorisation scheme ... by pleading the impossibility of carrying out more exhaustive prior examinations by reason of the considerable quantity of processing aids which may be used or by reason of the fact that manufacturing processes are constantly changing. As is apparent from Articles 6 and 7 of Regulation No 178/2002, concerning the analysis of risks and the application of the precautionary principle, such an approach does not correspond to the requirements laid down by the Community legislature as regards both Community and national food legislation and designed to achieve the general objective of a high level of health protection.[51]

Thus, the risk analysis methodology is interpreted by the Court in connection with a high level of health protection, requiring in-depth assessments of the risks and imposing high-scientific thresholds for national derogations to prove their necessity. A correct application of the precautionary principle also requires 'identification of the potentially negative consequences for health' and a 'comprehensive assessment of the risk to health based on the most reliable scientific data and the most recent results of international research'.[52] In other words, the Court examines the suitability and necessity of national measures taken under scientific uncertainty.

In the same judgment, however, the Court has reiterated that, to the extent that uncertainties exist in the current state of scientific research, it is for the Member States to determine the level of protection they deem appropriate.[53] If Member States have the power to set the level of protection, the third element of the proportionality principle, a test of the proportionality *stricto sensu*, seems to be excluded (Jans, 2000). The Court will not balance the protection of human health against free movement of goods in the EU.

The question of whether or not the Court applies the proportionality *stricto sensu* test to national food safety legislation is not that simple in practice. For example, it is widely accepted in case-law and policy documents that a zero risk is seldom a realistic objective for food safety risk management, even though EU and national measures pursue a high level of health protection. It implies that national measures cannot go beyond what is necessary to avoid a risk to human health – measures taken against risks that are very low cannot be justified even in light of the

---

[50] Dir. 88/344/EEC, OJ 1988, L 157/28.
[51] Case C-333/08, *supra* note 49, para. 103.
[52] Case C-333/08, *supra* note 49, para. 92.
[53] *Ibid.*, at para. 85. See also Van der Meulen (2010b).

precautionary principle.[54] National legislation is therefore not completely immune to balancing of interests in the application of the proportionality principle in the narrow sense. However, the Court will not carry out such a test profoundly, rather limiting itself to evident imbalances between the free movement of goods and benefits for human health.

The scrutiny in reviewing national legislation adopted under Article 114 TFEU, i.e. in areas which have been harmonised, is connected with the level of protection set out in the EU act. It is apparent that the proportionality of the national derogation cannot be judged in isolation from the existing EU harmonized legislation, which according to Article 114(3) TFEU, is supposed to set a high standard of protection, taking into account recent scientific developments. In this context, criteria for assessing the proportionality of national measures seem to be more stringent than those underlying Article 36 TFEU.[55] A Member State must not only prove that a derogating measure is necessary and does not constitute a means of arbitrary discrimination or a disguised restriction on trade, but also that the level of protection ensured by the harmonization measure is not sufficient in light of scientific facts and/or that the measure contains an error of fact or of assessment.[56]

In practice, this requirement could mean that Member States invoking Article 114(4) TFEU must still either prove that their population is in a specific situation of risk compared to the EU as a whole[57] or challenge the scientific risk assessment on which the EU measure is based to show that the harmonized legislation does not ensure sufficient protection. As discussed above, it was the specific situation in the Netherlands related to the use of creosote that was the basis for approval of the national derogation in 2002. Similar criteria were applied by the Commission in assessing the legality of Swedish measures concerning the use of certain colours and sweeteners in foodstuffs, where the national studies provided by the Swedish authorities were confronted with the evaluations of the Scientific Committee on Food (SCF), on which the EU harmonization measure was based.[58] The Commission concluded that the Swedish studies did not show that the colours in

---

[54] See, in this regard, Dec. 2009/726/EC concerning interim protection measures taken by France as regards milk products coming from a holding where a scrapie case is confirmed, OJ 2009, L 258/27, paras 22-25.

[55] See Opinion of A.G. Tizzano, *supra* note 8, para. 99.

[56] Case 3/00, *supra* note 10, para. 93.

[57] See Opinion of A.G. Tizzano, *supra* note 8. In this regard he states that '[i]f it were a problem common to all or the majority of Member States, it would presumably already have been resolved by the directive, but if that were not so, it would be necessary to verify whether the conditions for challenging the directive were met, given that the directive must ensure not just general protection but a high level of protection; in any case, the problem would be of a general nature and it is not therefore possible to understand why it should be resolved only for the fortunate citizens of a single meticulous Member State, to the detriment of the uniform application of the harmonized rules and hence of the functioning of the internal market' (para. 77).

[58] Dec. 1999/5/EC, *supra* note 7.

question posed any particular health problem for the Swedish population compared with other Member States and that all the risks highlighted in the national reports were sufficiently dealt with in the SCF reports. It may seem indeed difficult to demonstrate, to quote A.G. Tizzano, 'that a directive negotiated over a period of years can have overlooked information so crucial as to justify the derogation', and that 'this information (or even new scientific evidence) can then emerge in the extremely short period of time between approval of the directive and the request to maintain national derogations'.[59]

It is clear, in the context of the precautionary principle, that divergent EU and national risk assessments can create uncertainties *per se*. The Danish case concerning the use of nitrites and nitrates, however, illustrates that a national measure may be justified by a different interpretation of the same scientific evidence on which the EU measure is based, or even of parts of that risk assessment, which, in the view of the national authorities, provide sufficient basis to conclude that uncertainty as to a risk to human health exists.

The Danish measure did not refer to national studies, but to the opinions of the Scientific Committee on Food – the same opinions on which the contested EU measure was based. The national measure derogated from Directive 95/2/EC establishing a list of additives other than colours and sweeteners authorised in food.[60] The EU list was based on the scientific evaluations of the SCF. Before the entry into force of the directive, the Danish legislation had contained a list of permitted food additives, including the conditions of use of nitrates, nitrites and sulphites and the Danish authorities intended to maintain the national list after transposing the directive.

In the view of the Danish authorities, the EU harmonisation measure, permitting an excessive use of sulphites, nitrites and nitrates, did not take into consideration the recommendations of the SCF. The Commission did not approve the Danish legislation as not justified by the need to protect public health (sulphites) and excessive in relation to this aim (nitrites and nitrates).[61] Subsequently, the Danish government decided to challenge the Commission decision before the Court of Justice, giving the Court the opportunity to rule for the first time on the application against the Commission's refusal to authorize the maintenance of national derogations.[62]

The Commission claimed that the evidence submitted by Denmark contained neither new scientific evidence nor a proof that the Danish population was in a

---

[59] *Supra* note 8, para. 82.
[60] OJ 1995, L 61/1.
[61] Dec. 1999/830/EC, OJ 1999, L 329/1.
[62] Case 3/00, *supra* note 10.

specific situation of risk concerning the consumption of nitrites compared to the European population. The national derogation was based solely on a different interpretation of the scientific opinion given by the SCF, which – according to the Danish authorities – was not sufficiently taken into account in establishing an acceptable daily intake (ADI) for these additives.[63]

The Court, however, did not explicitly refer to the inadequacy of protection of the EU measure, but went on to state that, pursuant to Article 114(4) TFEU, a Member State may maintain in force national derogations on the basis of an assessment of risks made unilaterally by that Member State, if the measures are necessary to attain a level of protection that is higher than the harmonization measure:

> [T]he applicant Member State may, in order to justify maintaining such derogating national provisions, put forward the fact that its assessment of the risk to public health is different from that made by the Community legislature in the harmonization measure. In the light of the uncertainty inherent in assessing the public health risks ... divergent assessments of those risks can legitimately be made, without necessarily being based on new or different scientific evidence.[64]

The Court thus confirmed that Member States can decide on their acceptable level of risk, taking into account the same scientific evidence on which the EU risk management decision is based. The necessity of the national measure, however, must be justified in the light of the existing EU measure – therefore, the national authorities must show that the EU measure is insufficient to attain the objective set out by the national legislator.

To summarize, the very fact that national food safety measures, as exceptions to the free movement of goods, have to prove that their legislation is necessary to attain the objective of the protection of public health, requires a strong scientific basis. Hence, the necessity test is used a standard for judicial review of national measures, contrary to EU regulation where the legislator enjoys broader discretion. It is worth noting in this context that recourse to the precautionary principle justifying measures adopted under scientific uncertainty will – similarly – allow more room for discretion for precautionary measures adopted at EU level compared to national food safety regulation.

---

[63] *Ibid.*, para. 111. The SCF in its second opinion noted that the residual amounts of nitrites allowed by the directive were much higher than those to be expected from the addition justified by technological needs. The Danish government claimed that if the levels of nitrate residues were as high as those authorised by the harmonisation measure, the ADI would be exceeded.

[64] *ibid.*, para. 63.

5.5 Other legitimate factors in EU food safety regulation

5.5.1 Other legitimate factors lowering the level of health protection

In the case discussed above, the Danish authorities succeeded in proving that the national measures concerning the use of nitrites and nitrates as food additives were legitimate because the harmonization measure did not ensure sufficient protection of human health. In consequence, the Court annulled the Commission Decision which rejected the Danish provisions concerning the use of those additives.

According to Article 114(7) TFEU, when a Member State is authorised to maintain or introduce national provisions derogating from a harmonisation measure, the Commission is obliged to examine immediately whether to propose an adaptation to that measure. This is in line with the requirement that the Commission take a high level of protection as basis of its measures concerning human health, environmental or consumer protection, in particular taking account of any new development based on scientific facts.

The ruling delivered in March was discussed at the next meeting of the Standing Committee on the Food Chain and Animal Health in April 2003. The Commission also requested EFSA to give an urgent opinion on the minimum levels of nitrites and nitrates in meat products necessary to achieve the preservative effect and to guarantee microbiological safety and asked Member States to provide national data on the technological need and real amounts of those additives used in foods.

EFSA confirmed that lowering the maximum levels of nitrites and nitrates was necessary and that, instead of residual amounts, the ingoing amounts of nitrites and nitrates should be monitored (EFSA, 2003). Accordingly, the Commission proposed an amendment to the Directive, introducing the new method for establishing the permitted level of nitrites and nitrates. The proposal, however, allowed exemptions from this rule for several meat products from the UK, including Wiltshire bacon and ham, justified by their traditional method of production. For these products, the maximum residual level would be maintained.[65] Other Member States also asked for similar derogations for their traditional products, while some Member States were opposed to introducing specific exemptions based on factors other than scientific risk assessment.[66] In the amended directive, these exemptions were eventually included for several traditionally manufactured meat products

---

[65] According to the Commission, for these products, due to the nature of the manufacturing process, it was not possible to control the ingoing amount of curing salts absorbed by the meat.

[66] European Parliament, Draft Report on the proposal for a Directive amending Directive 95/2/EC on food additives other than colours and sweeteners and Directive 94/35/EC on sweeteners for use in foodstuffs (COM(2004)0650 – C6-0139/2004 – 2004/0237(COD)).

specifically named, as well as for products produced in a similar manner, which could if necessary be categorized in accordance with the comitology procedure.[67]

Despite the amendments to Directive 95/2 changing the authorizations for nitrites and nitrates as a follow up to the judgment of the Court of Justice and EFSA's opinion, Denmark once again decided to maintain, on the basis of Article 114(4) TFEU, its stricter national law. In November 2007 the Danish government notified the Commission of its intention not to transpose the amended provisions insofar as they concerned the addition of nitrite to meat products. In comparison with the EU measure, the Danish provisions laid down lower maximum added amounts for several types of meat products and did not allow exceptions referring to the traditional methods of production, forbidding the placing on the market of products for which no ingoing amounts could be established. Denmark highlighted that these provisions had been placed for many years and had proved to keep a very low rate of food poisoning cases caused by sausages compared to other Member States.

Like in the previous request to maintain the national measures, Denmark referred to the opinions of the EU scientific bodies. The Danish government considered the national provisions in full compliance with the recommendations of the SCF, as well as EFSA's risk assessment from 2003, since none of these opinions suggested any exceptions to the principle of controlling maximum 'added amounts' of nitrites rather than residual levels.[68]

In the light of the facts provided by Denmark and confirmed by EFSA, the Commission concluded that the Danish provisions were based on the scientific evidence of the SCF and EFSA and thus justified on grounds of protection of public health.[69] The decision was all the more easy when taking into account that the national derogations did not seem to constitute a serious obstacle to the functioning of the internal market. Denmark showed that imports of meat products from other Member States had even been increasing during the time when the stricter provisions were in force.

In this case, the Danish derogation was approved despite the fact that the Commission expressly stated that the harmonization measure was adequate to ensure the microbiological safety of meat products.[70] Although harmonized legislation has to ensure a high level of health protection, this level does not necessarily have to be the highest that is technically possible.[71] In its risk management decision,

---

[67] Dir. 2006/52/EC, OJ 2006, L 204/10.

[68] The Danish authorities highlighted the fact that the residual nitrite is not a reliable indicator because residual values may not reveal even very high additions of nitrites, which may in turn lead to a very high formation of nitrosamines.

[69] Dec. 2008/448/EC, OJ 2008, L 157/98. The approval was temporary – until 23 May 2010 (2 years).

[70] *Ibid.*, para. 41.

[71] Case T-13/99, *supra* note 41, para. 152.

the Commission decided not to pursue the strictest maximum levels for nitrites and took into account other factors – relating to specific traditional methods of production and the diversity of the meat products across the EU. Similarly, Regulation 178/2002 defines the aim and scope of food law as 'the assurance of a high level of protection of human health and consumers' interest in relation to food, taking into account in particular the diversity in the supply of food including traditional products, whilst ensuring the effective functioning of the internal market.'[72]

Although the General Food Law applies at both EU and national level, as we have shown in the previous cases, discretion permitted for the EU institutions and Member States varies. Mortelmans (2002), in his analysis of the relationship between the Treaty provisions and secondary legislation, argues that the EU institutions are bound by the Treaty rules relating to free movement in the same way as the Member States, but enjoy a broader margin of discretion. Referring to the risk analysis framework, in the case of conflict between the free movement rules, discretion is especially relevant to the extent to which other legitimate factors can constitute a basis in risk management decisions.

In the amended directive on the use of nitrites and nitrates as food additives, the EU legislature, in addition to the scientific opinion, took into account other legitimate factors in the risk management choice. This measure was easily challenged by the Member State requesting a derogation from the harmonisation measure and invoking the scientific opinion on which the harmonisation measure was based. The other legitimate factors, taking into account traditional methods of production and the diversity of products on the internal market, lowered the level of protection of the harmonization measure. This situation is clearly not possible in the case of national derogations based on Article 114(4) TFEU because they have to be more stringent than the EU harmonisation measure.

In line with Article 114 TFEU, one would expect the mechanism of the internal market to bring the differentiated standards to a uniform regulation in accordance with the 'race to the top'. The food safety paradigm, however, is influenced by socio-cultural implications of trade in foods. If the EU regulator decided to follow the scientific opinions, traditional cured meat products would disappear from the market. Thus, permitting exceptions justified by specific methods of production in this case does not seem to hinder the functioning of the internal market, even if several Member States decide to opt-out, along the lines marked by the scientific opinions of EFSA.

---

[72] Art. 1(1) GFL.

5.5.2 Other legitimate factors raising the level of health protection

Another possibility resulting from taking into consideration other values and interests in a regulatory procedure is a measure ensuring a higher level of protection compared to the recommendation of the scientific risk assessment on which it is based. This possibility in theory concerns both EU measures and national derogations. The extent to which other legitimate factors can be taken into account by decision makers, again, depends on the discretion framed by the Treaty provisions relating to the functioning of the internal market.

The EU ban on the use of growth promoting hormones except for therapeutic or other purposes and the ban on the import or intra-EU trade in meat from cattle treated with these hormones is an example of an EU measure introduced to a large extent as a response to consumer concerns about the safety of meat treated with hormones, prompted by abuse of good veterinary practice, rather than conclusive scientific studies, although these non-scientific assumptions were largely concealed behind scientific uncertainty.[73] The relationship between this measure and the internal market was dealt with in the *Fedesa* case, a preliminary ruling concerning proceedings brought before a national court in the UK, where the applicants challenged the validity of the national regulations implementing Directive 88/146/EEC.[74] The applicants argued that the directive banning the use of hormones in meat was not underpinned by scientific evidence and hence frustrated the legitimate expectations of traders.[75] The Court did not uphold this claim, referring to the discretionary power conferred on the EU institutions, and did not discuss the existence of scientific evidence showing the safety of the five hormones, merely stating that:

> [F]aced with divergent appraisals by the national authorities of the Member States, reflected in the differences between existing national legislation, the Council remained within the limits of its discretionary power in deciding to adopt the solution of prohibiting the hormones in question, and respond in that way to the concerns expressed by the European Parliament, the Economic and Social Committee and by several consumer organizations.[76]

---

[73] Dir. 81/602/EEC banned the use of substances having hormonal action, but the use of oestradiol-17β, progesterone, testosterone, trenbolone acetate and zeranol was left to be regulated in accordance with the individual regulatory schemes in Member States and other countries, pending further research. Dir. 85/649/EEC extended the ban to include these five hormones, but it was annulled by the Court of Justice on procedural grounds. The same provisions were re-introduced by Dir. 88/146/EEC. The ban was a reaction to 'hormone scandals' in Italy in the late 1970s. Premature development of schoolchildren was suspected to be linked to illegal hormones in meat from school canteens. See Roberts (1998).

[74] Case C-331/88, *Fedesa* [1990] ECR I-4023.

[75] *Ibid.*, para. 7.

[76] *Ibid.*, para. 9.

In the *Fedesa* case, it was the EU measure that hindered free movement of goods, and not the opposite – more typical – scenario, i.e. a Member State against, and the Commission for, the internal market.[77] The discretion in responding to consumer perceptions and public concerns is much wider than in the case of national measures. Consider the case *Land Oberösterreich* v. *Commission*, concerning the lawfulness of national measures banning genetic engineering. The Land Oberösterreich intended to prohibit the cultivation of genetically modified organisms (GMOs) and breeding of transgenic animals and notified its draft measures in accordance with Article 114(5) TFEU. The Commission rejected the request for derogation.[78] After the Court of First Instance (now renamed the General Court) dismissed the action brought by the Land Oberösterreich and Austria in which they sought annulment of the decision,[79] the applicants appealed to the European Court of Justice to set aside the judgment.[80]

Austria claimed that the ban was justified by the specificity of the province of Upper Austria, founded on the existence of small size farms and organic production in the area. The Land Oberösterreich provided a study suggesting that it is not possible for organic and conventional production to coexist alongside a vast GMO cultivation. Thus, the Austrian request for derogation related more to socio-economic aspects than to scientific evidence concerning the protection of the environment. The applicant stated itself that the notified measure was 'prompted by the fear of having to face the presence of GMOs'.[81]

The Commission based the rejection on the opinion from EFSA, in which it concluded that the information provided by Austria did not contain any new scientific evidence which could justify the measure. Moreover, in the view of the Commission, the small size of farms was a common characteristic for all Member States, and hence did not constitute a problem specific to Upper Austria. Finally, the Commission rejected Austria's argument justifying the national derogation by recourse to the precautionary principle as too general and lacking substance. In this case too the Commission referred to EFSA's opinion, which did not identify any risk that 'would justify taking action on the basis of the precautionary principle at Community or national level.'[82]

---

[77] Mortelmans refers in this regard to the *BSE* case, where the UK argued against the Commission measure prohibiting exports of UK beef (Case C-180/96, *UK* v. *Commission* [1998] ECR I-2288) (Mortelmans, 2002).

[78] Dec. 2003/653/EC, OJ 2003, L 230/24.

[79] Cases T-366/03 and T-235/04, *Land Oberösterreich* and *Austria* v. *Commission* [2005] ECR II-4005.

[80] Cases C-439/05 P and C-454/05 P, *Land Oberösterreich* and *Austria* v. *Commission* [2007] ECR I-7141.

[81] Cases T-366/03 and T-235/04, *supra* note 79, para. 67. See also Dec. 2008/62/EC, OJ 2008, L 16/17, concerning Polish provisions derogating from Dir. 2001/18/EC on the deliberate release into the environment of GMOs. The rejected national derogations referred to similar justifications, i.e. the need to fulfil the expectations of Polish society, a high level of fragmentation of agriculture and interest in organic production.

[82] Cases T-366/03 and T-235/04, *supra* note 79, para. 73.

The Court dismissed the appeals of Austria and the Land Oberösterreich. Obviously, the case concerned the lawfulness of national measures notified under Article 114(5) TFEU, which explicitly requires that a national derogation be based on new scientific evidence on the grounds of a problem specific to that Member State. However, the ruling would not be different if the national authorities wished to maintain existing national measures on the basis of Article 114(4) TFEU, which does not refer to these conditions. In fact, the autonomy of Member States is limited under both opt-out clauses – both possibilities of derogation are closely linked to the scientific evidence put forward by the Member State and the risk assessment on which EU harmonisation measures are based.

According to Article 114(4) TFEU, national derogations must be justified on grounds listed in Article 36 TFEU or relating to the environment or the working environment. The narrow interpretation of the exception relating to the protection of health and life of humans, on which food safety measures are based, requires that the national authorities must carry out a comprehensive risk assessment. As we have said, this in-depth, case-by-case assessment should appraise the degree of probability of harmful effects and the actual risk a foodstuff might pose for health.[83] This restrictive approach practically precludes Member States from invoking factors other than science, be they social, traditional or economic reasons, to justify their national derogations.

5.6 Conclusion

The risk analysis methodology established in Regulation 178/2002 applies to both EU and national food safety measures. The methodology, however, has different implications for the EU and national risk managers in the context of the Treaty provisions relating to the internal market and judicial review of measures protecting human health. This difference is reflected in the role science and other legitimate factors play in the risk management process.

The EU institutions enjoy a broad discretion in setting food safety standards, which can be largely motivated by other legitimate factors. The inclusion of these factors can lead to laxer standards, justified, e.g. by the diversity in the food supply or traditional methods of production, but also to stricter measures, based, e.g. on consumer risk perception.

Contrary to EU measures, national exceptions to the rule of the free movement of goods are dealt with restrictively. Because food safety regulations directly affect the functioning of the internal market, they are usually totally harmonized, depriving Member States of any powers to legislate, apart from the possibility of derogation provided in Article 114(4) TFEU. The national authorities must demonstrate that

---

[83] See Case C-333/08, *supra* note 49, para. 101.

their legislation is necessary to protect human health and, to this end, they must carry out a risk assessment. The internal market mechanism and the necessity to resist national protectionism leave practically no possibility to justify national food safety measures by factors other than science.

Moreover, in the case of diverging national and EU scientific opinions, the EU Courts tend to follow the scientific expertise provided by EU scientific institutions, with EFSA playing a central role. Thus, it seems rather difficult for a Member State to overturn EFSA's risk assessment on which an EU measure is based. Yet, for the same reason, Member States may rely on EFSA's opinion to justify maintaining stricter national provisions, either by referring to a different interpretation, e.g. highlighting uncertainties, or by challenging an EU risk management decision influenced by other legitimate factors and ensuring a level of protection lower than suggested in the scientific risk assessment ('race to the top').

The inclusion of socio-economic considerations in the decision making process creates an opportunity to promote diversity in the internal market by taking into account local traditional, cultural and social implications, as opposed to technocratic food safety regulation based on 'sound' science. Striking a balance between the safety paradigm and other legitimate factors at EU rather than national level is understandable from an internal market perspective, but raises concerns about the legitimacy of risk regulation. In addition to the questions of transparency and procedural guarantees for public participation in the decision-making process, the challenge facing food safety governance at the EU level involves establishing a credible risk governance across a variety of different national cultures.

This analysis leads to the conclusion that Member States have no possibility to invoke factors other than science to justify their food safety measures. EU harmonization measures, however, may explicitly allow Member States to adopt more specific provisions, taking into account particular situations or problems existing in their territories. This approach is already an important element of the so-called 'hygiene package' – the re-organized EU regulatory framework for food hygiene and safety, which became applicable in 2006. Flexibility is necessary to guarantee the viability of traditional products and small food producers. According to Article 10(3-8) of Regulation 853/2004 laying down specific hygiene rules for food of animal origin,[84] Member States may introduce national measures adapting the

---

[84] OJ 2004, L 226/22 (Corrigendum).

general hygiene requirements to traditional methods of production[85] or to food businesses in regions that are subject to geographical constraints. The national measures, however, cannot compromise the objectives of food safety. Products to which specific national hygiene measures apply are in free circulation in the internal market. The procedure is transparent: Member States have to notify their measures to the Commission and other Member States and any disagreement is discussed by the Standing Committee on the Food Chain and Animal Health.

In determining the content of harmonization measures the objective of the internal market is the most important. The wide use of the flexible method of harmonization, however, could touch on diversity of the internal market, by giving the Member States the possibility to apply adaptations, taking into account local factors. This method of harmonization creates an opportunity for incorporating a wider set of societal concerns into the decision making process at the national level. The flexible approach to harmonization, giving more possibilities to include non-scientific factors in national measures, as well as a clear formulation of how science and a wide set of other legitimate factors interact in food safety policy-making, could improve the credibility of risk governance in this field.

---

[85] 'Foods with traditional characteristics' are defined in Art. 7 Reg. 2074/2005 laying down implementing measures for certain products under Reg. 853/2004 as 'foods that, in the Member State in which they are traditionally manufactured, are:

a. recognized historically as traditional products; or
b. manufactured according to codified or registered technical references to the traditional process, or according to traditional production methods; or
c. protected as traditional food products by a Community, national, regional or local law' (OJ 2005, L 338/27).

6. Conclusions

6.1 Main findings

Two opposing dynamics are driving regulatory activities at national, EU and international levels. On the one hand, economic integration and the reduction of obstacles to trade are progressing regionally and globally. On the other hand, a body of law that interferes with market processes to protect public health or the environment – referred to as risk regulation – is becoming a prominent form of regulation in the EU.

The objective of this study was to analyse the concepts of risk analysis and the precautionary principle established in the General Food Law and their impact on food safety regulation in EU multi-level governance. These general principles introduce a comprehensive model for the decision-making process in the area of risk regulation pertaining to food safety. The main findings relating to the role of risk analysis and the precautionary principle in EU food law presented below are grouped around the four research questions formulated at the beginning of the study.

6.1.1 What is the constitutional framework for risk analysis?

Food law is almost entirely centralised at EU level. Because the internal market is an area where the Union and the Member States share competence (Article 4(2) TFEU), the Member States can exercise their competence only to the extent that the EU has not acted. Because food safety is inextricably related to trade, measures taken by a Member State are likely to affect the free movement of goods. Therefore, to achieve the internal market objectives, the EU has gradually taken over the national competences in this area.

Once laws have been harmonised, the possibility of Member States to derogate from EU measures in the field of food safety is limited to the exception provided in Article 114(4) TFEU or to safeguard clauses included in the harmonisation measure itself. Safeguard clauses allow Member States to temporarily suspend trade of a food if they consider it dangerous to human health. The exception in Article 114(4) TFEU gives Member States the possibility to maintain – with the Commission's approval – stricter national standards. As has been shown in this study, ultimately harmonisation measures are usually amended in line with the stricter national standard, which becomes a common standard for the whole EU. The national exceptions have developed into a vehicle for a 'race to the top', where the common standard is determined by scientific risk assessment.

These patterns in the functioning of the internal market lead us to analyse the role of science in food safety law. In the EU, food safety is the only area of risk regulation where a comprehensive risk analysis model including the precautionary principle has been codified into general principles governing policy. These principles apply to both EU and national measures. In deciding on food safety issues, decision-makers rely on technical and scientific expertise. Risk analysis is a methodology incorporating scientific evidence into the decision-making process: regulatory decisions are based on risk assessment provided by experts. If risk assessment is inconclusive, temporary measures may be introduced based on the precautionary principle.

In risk regulation scientific evidence (showing risk or at least scientific uncertainty regarding risk) occupies a central position. This holds all the more true for risk regulation closely connected to the internal market, where science is used to draw a demarcation line between the protection of human health and trade objectives. The scientific food safety governance established by the General Food Law did not vest the European Food Safety Authority with a decisive voice in case of conflicting national scientific opinions. Nevertheless, EFSA was designed to become a scientific authority for the entire EU. It is increasingly becoming a reference point for the European Courts, which choose to rely on EFSA's opinions rather than on scientific evidence from other sources. Moreover, the 'scientific' mediation procedure set out in Article 60 GFL, as well as EFSA's obligation to cooperate with national bodies in case of diverging scientific opinions, have a potential to become important elements of the EU internal market. While mutual recognition is the EU answer to obstacles to the free movement of goods created by different national technical and quality standards for food, close cooperation and integration of scientific expertise within the EU (*'mutual scientific opinion'*) can contribute to eliminating barriers to intra-EU trade in food.

In this way, the constitutional framework for risk analysis consists of the Treaty provisions on the internal market, including the possibilities to derogate in conjunction with the principles requiring and shaping the procedure for scientific underpinning of risk management decisions.

6.1.2 How are other legitimate factors incorporated into the decision-making process at EU and national levels of food safety governance?

The science-based food law translated into the internal market language means that a Member State must provide scientific evidence if it wants to maintain a stricter national standard or, in a non-harmonised area, to restrict or ban a product from its territory on the basis of Article 36 TFEU. The national measure must be proportional and necessary for the desired level of protection. The principle of the free movement of goods, however, does not bind the EU institutions in the same way as the national legislators. Unlike the national legislators, the EU institutions enjoy a broad discretion in setting food safety standards, which can be largely motivated by factors other than science. The difference between the discretion the EU institutions and national authorities enjoy in the field of food safety was a key element of this study.

The General Food Law recognises that scientific risk assessment alone cannot provide all the information on which a risk management decision should be based, and that *other factors legitimate* to the matter under consideration should be taken into account. These factors include, e.g. societal values and preferences, and economic, traditional, ethical, or environmental considerations. Although the role of other legitimate factors is not clearly defined, the extent to which food safety measures can deviate from scientific evaluations in considering them depends on how much discretion is conferred on the decision-makers.

The discretion is coupled with the standard of review (the amount of deference) applied by the courts to assess measures. In the EU, the proportionality test applied by the judiciary is usually defined as three steps: rationality test, necessity test, and proportionality *stricto sensu*. We found that, in judicial review of measures taken at EU level, the European Courts do not go beyond the first element of the proportionality test – the rationality review. Contrary to EU measures, national exceptions are dealt with restrictively. Courts apply the necessity test, which means that national food safety measures must be based on the findings of risk assessment showing a risk to human health. Even though the last element of the proportionality test – proportionality *stricto sensu* – is not likely to be invoked, the internal market mechanism still practically precludes the possibility to maintain national food safety measures based on factors other than scientific evidence.

Hence, although the risk analysis methodology applies to both EU and national measures, it has different implications for the EU and national risk decision makers. The EU institutions enjoy a broad discretion in setting food safety standards, which can be largely motivated by other legitimate factors. The inclusion of these factors at EU level can lead to not only laxer standards, justified, e.g. by the need to maintain the diversity in the food supply or traditional methods of production, but also to stricter measures, based on, e.g. consumer risk perception (see Figure 6.1). The national measures cannot go beyond the findings of risk assessment.

In a picture that emerges from the analysis of the discretion of risk managers at two levels of the EU, food safety governance homogeneity of the internal market[1] and risk-based approach are clearly brought to the forefront. Because Member States cannot base their food safety measures on considerations other than science, paradoxically deciding on societal preferences and local and traditional factors can only take place at EU level. This is understandable from an internal market perspective, but raises concerns about the legitimacy of food safety regulation in the EU.

---

[1] I.e. no or limited differences between national laws.

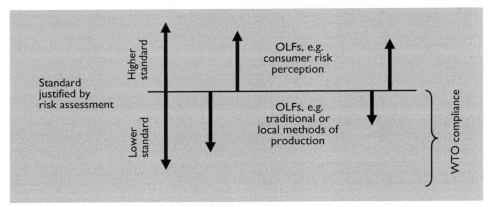

Figure 6.1. EU food safety standards influenced by 'other legitimate factors' (OLFs).

Therefore, a more flexible and inclusive approach to harmonisation is recommended. Harmonisation measures based on such an approach should contain explicit provisions enabling national legislators to derogate, taking into account national specificities. The possibility to include some non-scientific factors at national level would better accommodate the European patchwork of diverse cultures and traditions relating to food in the EU internal market. Some legislative acts already contain rules giving the national legislator the possibility to derogate for these reasons. The need for a change towards a more flexible approach is recognised in some other areas, for example in legislation on GMOs.

From a market perspective, the inclusion of other legitimate factors lowering the level of protection compared to scientific advice is less harmful to trade than risk management measures resulting in a level of protection higher than recommended by risk assessment. An EU measure ensuring a level of protection lower than recommended in risk assessment may trigger in some cases a race to the top if one or more Member States will be allowed to maintain their stricter national standards, but will not lead to a barrier in international trade. In the scenario where a measure is more trade restrictive than results from risk evaluation, not only are the national risk managers constrained in their choices, but also the EU decision-makers: by the international trade regime established by the WTO Agreements.

The WTO Agreement on the Application of Sanitary and Phytosanitary Measures (SPS Agreement) stipulates that all measures aimed at the protection of human health that hinder international trade must be based on scientific evidence or, in cases of scientific uncertainty, provisionally adopted, pending additional information for a more objective risk assessment and reviewed accordingly.[2]

---

[2] Art. 5.7 SPS.

Food safety measures that restrict trade must be preceded by the evaluation of available scientific information, and recourse to the concept of precaution can only take place within this decision-making paradigm.

Even though the General Food Law provides 'other legitimate factors' as an element in the risk analysis framework, this element can be used only by the EU institutions, not by national authorities. This wider scope for the EU institutions may, however, not be compatible with WTO law.

6.1.3 What is the scope of application of the precautionary principle as defined in Article 7 of the General Food Law?

Food safety measures risk non-compliance with the WTO regime if they are not underpinned by scientific evidence. In this sense, the SPS Agreement defines boundaries for EU food safety law. In cases where the EU institutions desire to go above the level of protection recommended by risk assessment, they try to widen the WTO boundaries by stretching out the concept of the precautionary principle. The precautionary principle applies when risk assessment is inconclusive. In the EU approach, the precautionary principle is a risk management tool in the sense that the decision whether the scientific evidence is conclusive or not is left to risk managers, and not to scientists. In consequence, uncertainty becomes politicised and public risk perception, as well as other factors, be they economic or traditional, creep into the risk analysis procedure under the disguise of the precautionary principle.

The General Food Law contains a definition of the precautionary principle. This is the first time a legal definition of the principle was formulated, which applies to food safety measures both at EU and national levels. Even the EU definition, however, has difficulties to accommodate the wide range of EU measures referring to this 'principle'. Some applications in the field of food safety, built up throughout the evolution of the principle from a general directing principle to more concrete legal shapes, although undoubtedly having 'protective' or 'precautionary' character, do not meet the criteria set out in the definition.

Paradoxically, this terminological confusion resulting from the application of the precautionary principle that is different from the conditions set out in the definition concerns some of the most celebrated examples of its application. Legislative frameworks setting out prior authorisation schemes used to be seen as 'precautionary' measures. These schemes set up a legislative framework, under which regulatory decisions concerning the placing on the market of certain categories of foodstuffs are being issued. Under this framework, the burden of producing scientific evidence is shifted from authorities acting on behalf of society to the proponent of the placing of a product on the market. Some categories of foods, such as genetically modified foods or novel foods, are thus considered a priori hazardous, until – to the satisfaction of the authorities – the producer or importer proves otherwise.

Undoubtedly, prior authorisation schemes are aimed at ensuring a high level of health protection and clearly have a precautionary character. They do not, however, meet the conditions laid down in Article 7 GFL. The mere presumption of risk from unknown products cannot substitute for 'scientific uncertainty' following assessment of available information. Similarly, the definition does not mention a shifting of a burden of producing scientific evidence as a consequence of the application of the principle – on the contrary, it allows authorities to act. This and other examples of measures based on the precautionary principle that do not meet the requirements set out in the definition confirm that the scope of the precautionary principle as codified in the General Food Law is narrower than previously used by the EU decision-makers (and applied by EU Courts). This requires rethinking both in light of the general principles of the General Food Law and the rationale behind prior authorisation schemes.

The scope of the precautionary principle as codified in Article 7 GFL is limited to empowering authorities to take risk management measures in case of inconclusive risk assessment. It does not justify action in the absence of risk assessment, nor does it reverse the burden of scientific substantiation.

6.1.4 Do the risk analysis methodology and the precautionary principle apply only to implementing regulatory measures or do they set limitations on the legislator as well?

Ultimately, the fundamental question this study set out to answer concerned the very scope of application of the precautionary principle and risk analysis. Risk analysis is usually referred to the setting of specific food safety standards or, in the context of pre-market approvals, to authorisations of individual products. The adoption of detailed food safety regulation referring to technical or scientific knowledge to a large extent takes place within the realm of public administration. We analysed whether the requirement of science-based food safety measures applies to basic legislative acts as well.

The question was answered by putting the EU prior authorisation scheme concerning novel foods to two tests: the interpretation of risk analysis and the precautionary principle by the European Court of Justice in a recent case concerning *national* prior authorisation schemes for food processing aids, and from an international perspective – by analysing the meta-framework set up by the WTO SPS Agreement.

In the case concerning the national law, the Court stated that the principles of food law set out in Articles 6 and 7 of the General Food Law require that food safety measures should normally be underpinned by an assessment of risks posed by a product or category of products. In consequence, prior authorisation schemes introduced without scientific reason do not correspond to this requirement. While the case concerned national law, the scope of the General Food Law explicitly covers national and EU law equally.

Similarly, the analysis of the WTO Agreement on the Application of Sanitary and Phytosanitary Measures (SPS Agreement) showed that, although allowing prior authorisations schemes in general, the SPS Agreement does not distinguish between basic and implementing acts: all SPS measures must be based on scientific evidence or – in cases of scientific uncertainty – on Article 5.7 SPS.

Under both WTO and EU laws, all food safety measures, whether legislative or administrative, that restrict trade must be preceded by the evaluation of available scientific information, and if they refer to the concept of precaution, they must refer to the possibility of harmful effects identified by the assessment of available information. If a measure is more trade-restrictive than the result of risk assessment indicates, it risks non-compliance with the decision-making paradigm under the WTO regime. This also refers to precautionary measures based on considerations other than science.[3]

In this way, the study has gone full circle and come back to the title of this book. The objective of the study was to analyse the concepts of risk analysis and the precautionary principle established in the General Food Law and their impact on food safety regulation in EU multi-level governance. We have shown that the general principles of food law introduce a comprehensive model for the decision-making process in this field. Their role is therefore to *regulate* food safety *law*.

6.2 Areas for future research

Although the emerging law of risk regulation at both EU and international levels is gradually gaining ground, the risk analysis methodology has so far received little attention in legal literature. Rather, the concept is evaluated as a general policy guidance with the question of how public authorities should deal with uncertainties as the core issue (De Vries *et al.*, 2011; Klint Jensen, 2006; Miller, 1999; Rogers, 2001; Van Asselt and Vos, 2005), or in the context of controversies surrounding the issue of judicial review of measures based on scientific evidence (Christoforou, 2000; Dąbrowska, 2004; Harlow, 2004; Hervey, 2001; McNelis, 2001; Peel, 2004). The added value of this study consisted in applying the general principles of food law to 'real life'. Procedures of incorporating scientific risk assessment, scientific uncertainty and other legitimate factors into the decision-making process were considered within the Treaty framework and the internal market mechanism, as well as secondary legislation, national laws, case-law,

---

[3] Cf. Goldstein and Carruth (2004).

and WTO obligations. Drawing on the results of this study, topics outlined below create paths that promise to make innovative contributions to our understanding of how risks should be regulated.

In line with the key elements of the governance of risk regulation – openness, effectiveness, participation, accountability, and coherence of policies – this study points in the following directions for future research:
- Connecting the risk analysis methodology with the regulatory reform in the EU known as 'Better Regulation' – to improve openness and accountability of risk analysis.
- Analysing possible implications of the WTO Agreements on the discretion of the EU legislature – with a view to increase the coherence of food safety policy and the protection of legitimate expectations.

In environmental law, the Aarhus Convention provides for the right of everyone to receive environmental information that is held by public authorities. This procedure includes information on policies and measures taken, but also the right to participate in environmental decision-making by commenting on proposals for measures relating to the environment. In food safety, although the main goals of the 2002 policy reform were to ensure openness and transparency, the EU remains a black box. No link between the scientific opinions of EFSA and the Commission's proposals in the field of food safety exists and no reference is made to the risk analysis methodology in documents accompanying legislative proposals: how scientific evidence was incorporated into the process and what other factors were taken into account.

The broad discretion at the EU level of food safety governance resulting in non-compliance with the international trade law, in connection with the black box of the EU decision-making process, is criticized. No reference to scientific arguments underpinning measures restricting trade frustrates the legitimate expectations of traders and makes it extremely difficult (or impossible) to challenge the legality of EU measures before the courts. This discretion and the lax approach to the risk analysis methodology have also raised questions about compliance of EU law with the SPS Agreement (Slotboom, 1999).

Future research could focus on opening the 'black box' of the risk analysis methodology, by tackling the following issues:
- What real implications should the risk analysis methodology have on the policy-making processes in the EU?
- What impact should risk analysis have on the discretion of EU decision-makers?
- Can risk analysis enhance the validity of EU risk regulation with respect to obligations related to WTO law and EU free movement of goods?

The risk analysis methodology could be analysed in the context of two meta-frameworks applied to EU law-making:

1. Meta-framework: Better Regulation (Impact Assessment)

    Regulatory reform measures known as 'Better Regulation' (BR) have recently been gaining increasing attention across the EU (EC 2002a; 2009). This measure is concerned with the process through which laws and delegated rule-making are formulated, appraised and enforced in the EU. Impact Assessment is the most popular instrument of BR, used to prepare evidence for decision makers. Emerging literature on BR compares this set of regulatory requirements to a meta-framework (Alemanno, 2009; Meuwese, 2008; Radaelli and Meuwese, 2010).

    As applied to risk governance and linked to risk analysis requirements, Impact Assessment may provide a new tool for putting the principles of transparency and openness in regulating risks into practice, by the obligation to explain scientific reasons and other legitimate factors underpinning risk measures, such as economic or social considerations. This innovative approach would allow incorporating the risk analysis methodology into the decision-making process. In the analysis of possible links between risk analysis and the BR meta-framework, a number of questions could be addressed, e.g. what is the legal status of BR requirements, can EU Courts review these requirements, or can private parties challenge impact assessments.

2. Meta-framework: international trade law

    International trade law, especially the SPS Agreement, touches upon the basic functions and mission of governments by setting limitations on their regulatory competences in fields of risk regulation in order to pursue a liberal international trade framework (Knodt, 2004; Skogstad, 2001; Slotboom, 1999, 2003). Such limitations raise criticism about the ability of national governments to set the level of protection of public health that *their citizens*, and not only experts, consider appropriate (Brom, 2004; Howse, 2000; Joerges and Neyer, 2003).

    Unlike other international agreements concluded by the EU, the EU Courts do not consider the WTO Agreements among the rules in light of which the legality of measures adopted by the EU is reviewed. This stance has triggered criticism in literature (Bronckers and Soopremanien, 2008; Kuijper and Bronckers, 2005; Zonnekeyn, 2004).

    Two important exceptions to this line of jurisprudence have been recognised by EU Courts. GATT/WTO provisions have the effect of binding the EU where it implements a particular obligation (*Nakajima* exception),[4] or where an EU measure refers expressly to specific GATT/WTO provisions (*Fediol* exception).[5] The *Nakajima/Fediol* doctrine might have important implications for EU food

---

[4] Case C-69/89, *Nakajima All Precision Co. Ltd* v. *Council*, [1991] ECR-2069, para. 31.

[5] Case 70/87, *Fédération de l'industrie de l'huilerie de la CEE (Fediol)* v. *Commission*, [1989] ECR 1781, paras 19-22.

safety measures, which often explicitly refer to the SPS Agreement and hence they should, in principle, meet this standard.[6]

Possible changes in the standard review of EU measures and their implications on the discretion of EU legislature in addressing the conflicting values of the protection of public health, food safety and the environment and trade rules could thus be assessed from the perspective of the WTO Agreements.

To be sure, there is no reason why such research should be confined to food safety regulation. An extension of the approach used in this study to other areas of risk regulation generates a research programme based on a comparative analysis, e.g. with environmental law. The risk analysis process will be different depending of the area of application, e.g. scientific evidence underpinning legislation on water pollution, which has less direct effects on the functioning of the internal market, will have a different function than a measure banning the use of a pesticide or authorising a novel food technology on the market. Examples of environmental and food safety measures could therefore allow to compare regulatory discretion and law-making procedures incorporating scientific evidence in the context of free trade rules. Moreover, environmental law provides an interesting perspective for food law because some principles applied to food law – like the controversial precautionary principle – originate in environmental law. Needless to say, comparative research could also serve as a 'test' of the validity of the approach presented here.

---

[6] An example of the 'WTO consciousness' is reflected, e.g. in the EU Regulation on hygiene in foodstuffs – Recital 18 states that the Regulation takes into account international obligations laid down in the SPS Agreement (Reg. 852/2004, OJ 2004, L 226/3).

Appendix: Articles 34, 36 and 114 TFEU

ARTICLE 34

(ex Article 28 EC)

Quantitative restrictions on imports and all measures having equivalent effect shall be prohibited between Member States.

ARTICLE 36

(ex Article 30 EC)

The provisions of Articles 34 and 35 shall not preclude prohibitions or restrictions on imports, exports or goods in transit justified on grounds of public morality, public policy or public security; the protection of health and life of humans, animals or plants; the protection of national treasures possessing artistic, historic or archaeological value; or the protection of industrial and commercial property. Such prohibitions or restrictions shall not, however, constitute a means of arbitrary discrimination or a disguised restriction on trade between Member States.

ARTICLE 114

(ex Article 95 EC)

1. Save where otherwise provided in the Treaties, the following provisions shall apply for the achievement of the objectives set out in Article 26. The European Parliament and the Council shall, acting in accordance with the ordinary legislative procedure and after consulting the Economic and Social Committee, adopt the measures for the approximation of the provisions laid down by law, regulation or administrative action in Member States which have as their object the establishment and functioning of the internal market.
2. Paragraph 1 shall not apply to fiscal provisions, to those relating to the free movement of persons nor to those relating to the rights and interests of employed persons.
3. The Commission, in its proposals envisaged in paragraph 1 concerning health, safety, environmental protection and consumer protection, will take as a base a high level of protection, taking account in particular of any new development based on scientific facts. Within their respective powers, the European Parliament and the Council will also seek to achieve this objective.
4. If, after the adoption of a harmonisation measure by the European Parliament and the Council, by the Council or by the Commission, a Member State deems it necessary to maintain national provisions on grounds of major needs referred to in Article 36, or relating to the protection of the environment or the working environment, it shall notify the Commission of these provisions as well as the grounds for maintaining them.
5. Moreover, without prejudice to paragraph 4, if, after the adoption of a harmonisation measure by the European Parliament and the Council, by the Council or by the Commission, a Member State deems it necessary to introduce national provisions based on new scientific evidence relating to the protection

of the environment or the working environment on grounds of a problem specific to that Member State arising after the adoption of the harmonisation measure, it shall notify the Commission of the envisaged provisions as well as the grounds for introducing them.

6. The Commission shall, within six months of the notifications as referred to in paragraphs 4 and 5, approve or reject the national provisions involved after having verified whether or not they are a means of arbitrary discrimination or a disguised restriction on trade between Member States and whether or not they shall constitute an obstacle to the functioning of the internal market.

   In the absence of a decision by the Commission within this period the national provisions referred to in paragraphs 4 and 5 shall be deemed to have been approved.

   When justified by the complexity of the matter and in the absence of danger for human health, the Commission may notify the Member State concerned that the period referred to in this paragraph may be extended for a further period of up to six months.

7. When, pursuant to paragraph 6, a Member State is authorised to maintain or introduce national provisions derogating from a harmonisation measure, the Commission shall immediately examine whether to propose an adaptation to that measure.

8. When a Member State raises a specific problem on public health in a field which has been the subject of prior harmonisation measures, it shall bring it to the attention of the Commission which shall immediately examine whether to propose appropriate measures to the Council.

9. By way of derogation from the procedure laid down in Articles 258 and 259, the Commission and any Member State may bring the matter directly before the Court of Justice of the European Union if it considers that another Member State is making improper use of the powers provided for in this Article.

10. The harmonisation measures referred to above shall, in appropriate cases, include a safeguard clause authorising the Member States to take, for one or more of the non-economic reasons referred to in Article 36, provisional measures subject to a Union control procedure.

References

AIRMIC, ALARM and IRM, 2002. A risk management standard. Available at: http://www. theirm.org/publications/documents/Risk_Management_Standard_030820.pdf.

Alemanno, A., 2007. Trade in food: regulatory and judicial approaches in the EC and the WTO. Cameron May, Cambridge, UK.

Alemanno, A., 2008. The European Food Safety Authority at five. European Food and Feed Law Review 1: 2-24.

Alemanno, A., 2009. The Better Regulation initiative at the judicial gate: a Trojan horse within the Commission's walls or the way forward? European Law Journal 15: 382-400.

Allio, L., Ballantine, B. and Meads, R., 2006. Enhancing the role of science in the decision-making of the European Union. Regulatory Toxicology and Pharmacology 44: 4-13.

American Chemical Society, 1998. Understanding risk analysis: a short guide for health, safety, and environmental policy making, Washington, DC, USA.

Ansell, C. and Vogel, D. (eds.), 2006a. What's the beef? the contested governance of European Food safety. The MIT Press, Cambridge, MA, USA.

Ansell, C. and Vogel, D., 2006b. The contested governance of European food safety regulation. In: C. Ansell and D. Vogel (eds.), What's the beef? the contested governance of European food safety. The MIT Press, Cambridge, MA, USA, pp. 3-32.

Beck, U., 1992. Risk society: towards a new modernity. Sage, London, UK.

Belvèze, H., 2003. Le principe de précaution et ses implications juridiques dans le domaine de la sécurité sanitaire des aliments. Revue Scientifique et Technique/Office International des Epizooties 22: 387-396.

Berends, G. and Carreño, I., 2005. Safeguards in food law – ensuring food scares are scarce. European Law Review 30: 386-405.

Bernstein, P.L., 1998. Against the gods: the remarkable story of risk. John Wiley and Sons, New York, NY, USA.

Bodansky, D., 1994. The precautionary principle in US environmental law. In: T. O'Riordan and J. Cameron (eds.), Interpreting the precautionary principle. Earthscan Publications, London, UK, pp. 203-228.

Brom, F.W.A., 2004. WTO, public reason and food public reasoning in the 'trade conflict' on GM-food. Ethical Theory and Moral Practice 7.

Bronckers, M. and Soopremanien, R., 2008. The impact of WTO law on European food regulation. European Food and Feed Law Review 6: 361-375.

Brookes, G., 2007. Economic impact assessment of the way in which the EU novel foods regulatory approval procedures affect the EU food sector. Briefing paper. For the Confederation of the Food and Drink Industries of the European Union (CIAA) & the Platform for Ingredients in Europe (PIE).

BSE Enquiry, 2000. The report. Available at: http://www.regulation.org.uk/bse.shtml.

Busch, L., 2004. Grades and standards in the social construction of safe food. In: M.E. Lien and B. Nerlich (eds.), The politics of food. Berg, Oxford, UK, pp. 163-178.

Byrne, D., 2002. EFSA: excellence, integrity and openness. Inaugural meeting of the Management Board of the European Food Safety Authority, 18 September 2002, Brussels, Belgium.

Caporaso, J.A., 1996. The European Union and forms of state: Westphalian, regulatory or post-modern? Journal of Common Market Studies 34: 29-52.

Cassese, S. (ed.), 2002. Per un'autorità nazionale della sicurezza alimentare. Coop-Il Sole 24 Ore. Milan, Italy.

Chalmers, D., 2003. 'Food for thought': reconciling European risks and traditional ways of life. Modern Law Review 66: 532-562.

Chiti, E., 2000. The emergence of a community administration: the case of European agencies. Common Market Law Review 37: 309-343.

Chiti, E., 2002. Le agenzie europee. Unita e decentramento nelle amministrazioni comunitarie. Cedam, Podova, Italy.

Christoforou, T., 2000. Settlement of science-based trade disputes in the WTO: a critical review of the developing case law in the face of scientific uncertainty. New York University Environmental Law Journal VIII: 622-648.

Christoforou, T., 2003. The precautionary principle and democratizing expertise: a European legal perspective. Science and Public Policy 30: 205-211.

Codex Alimentarius Commission (CAC), 1999. Report of the fourteenth session of the Codex Committee on General Principles. Paris, 19-23 April 1999. FAO/WHO, Rome, Italy.

Codex Alimentarius Commission (CAC), 2007a. Procedural manual 17th edition. FAO/WHO, Rome, Italy.

Codex Alimentarius Commission (CAC), 2007b. Working principles for risk analysis for food safety for application by governments. CAC/GL 62-2007. FAO/WHO, Rome, Italy.

Codex Alimentarius Commission (CAC), 2011. Procedural manual 20th edition. FAO/WHO, Rome, Italy.

Costato, L., 2003. Note introduttive. In: Istituto di Diritto Agrario Internazionale e Comparato (ed.), Le nuove leggi civili commentate: la sicurezza alimentare nell'Unione Europea Vol. 1-2. CEDAM, Padua, Italy.

Da Cruz Vilaça, J.L., 2004. The precautionary principle in EC Law. European Public Law 10: 369-406.

Dąbrowska, P., 2004. Risk, precaution and the internal market: who won the day in the recent *Monsanto* judgment of the European Court of Justice on GM foods? German Law Journal 5: 151-165.

De Búrca, G., 2000. Proportionality and subsidiarity as general principles of law. In: U. Bernitz and J. Nergelius (eds.), The general principles of European Community law. Kluwer Academic Publishing, Dordrecht, the Netherlands, pp. 95-112.

De Marchi, B., 2003. Public participation and risk governance. Science and Public Policy 30: 171-176.

De Sadeleer, N., 2002. Environmental Principles: from political slogans to legal rules. Oxford University Press, Oxford, UK.

De Sadeleer, N., 2003. Procedures for derogation from the principle of approximation of laws under Article 95 EC. Common Market Law Review 40: 889-915.

De Sadeleer, N., 2006. The precautionary principle in EC health and environment law. European Law Journal 12: 139-172.

De Sadeleer, N., 2010. Environnement et marché intérieur. Editions de l'Université de Bruxelles, Brussels, Belgium.

De Vries, G., Verhoeven, I. and Boeckhout, M., 2011. Taming uncertainty: the WRR approach to risk governance. Journal of Risk Research 14: 485-499.

Dehousse, R., 1997. Regolazione attraverso reti nella Comunita europea: il ruolo delle agenzie europee. Rivista Italiana di Diritto Pubblico Comunitario 1997: 629-650.

Dougan, M., 2000. Minimum harmonisation and the internal market. Common Market Law Review 37: 853-885.

Dressel, K., Böschen, S., Schneider, M., Viehöver, W., Wastian, M. and Wendler, F., 2006. Food safety regulation in Germany. In: E. Vos and F. Wendler (eds.), Food safety regulation in Europe: a comparative institutional analysis. Intersentia, Antwerp, Belgium, pp. 287-330.

Dutheil de la Rochère, J., 2007. Le principe de précaution In: J.-B. Auby and J. Dutheil de la Rochère (eds.), Droit Administratif Européen. Bruylant, Brussels, pp. 459-471.

Emiliou, N., 1996. The principle of proportionality in European law: a comparative study. Kluwer Law International, London, UK.

Eurobarometer, 2010. Special Eurobarometer 354 'food-related risks'. Available at: http://ec.europa.eu/public_opinion/archives/ebs/ebs_354_sum_en.pdf.

European Commission (EC), 1985a. Completing the internal market. White Paper from the Commission to the European Council (Milan, 28-29 June 1985). European Commission COM (85) 310 final. Brussels, Belgium.

European Commission (EC), 1985b. Completion of the internal market: community legislation on foodstuffs. European Commission COM (85) 603 final. Brussels, Belgium.

European Commission (EC), 1993. Commission report to the European Council on the adaptation of Community legislation to the subsidiarity principle. European Commission COM (93) 545 final. Brussels, Belgium.

European Commission (EC), 1997a. Communication from the Commission on Consumer Health and Food Safety. European Commission COM (97) 183 final. Brussels, Belgium.

European Commission (EC), 1997b. The general principles of Food Law in the European Union. Commission Green Paper. European Commission COM (97) 176 final. Brussels, Belgium.

European Commission (EC), 2000a. White paper on food safety. European Commission COM (1999) 719 final. Brussels, Belgium.

European Commission (EC), 2000b. Communication from the Commission on the precautionary principle. European Commission COM (2000) 1. Brussels, Belgium.

European Commission (EC), 2001a. European governance: a white paper. European Commission COM (2001) 428. Brussels, Belgium.

European Commission (EC), 2001b. Report of the working group 'Democratising expertise and establishing scientific reference systems' (Group 1b). White Paper on Governance, Work Area 1 – Broadening and enriching the public debate on European matters. Brussels, Belgium.

European Commission (EC), 2002a. European governance: better lawmaking. European Commission COM(2002) 275 final. Brussels, Belgium.

European Commission (EC), 2002b. Communication from the Commission concerning Article 95 (paragraphs 4, 5 and 6) of the Treaty Establishing the European Community. European Commission COM (2002) 760. Brussels, Belgium.

European Commission (EC), 2008. Draft report on impact assessment for a Regulation replacing Regulation (EC) No 258/97 on novel foods and novel food ingredients, COM(2007) 872 final. Brussels, Belgium.

European Commission (EC), 2009. Impact assessment guidelines. European Commission SEC(2009) 92. Brussels, Belgium.

European Commission (EC), 2010. Guidance on the implementation of Articles 11, 12, 14, 17, 18, 19 and 20 of Regulation (EC) No 178/2002 on General Food Law. Conclusions of the Standing Committee on the Food Chain and Animal Health, 26 January 2010. Brussels, Belgium. Available at: http://ec.europa.eu/food/food/foodlaw/guidance/docs/guidance_rev_8_en.pdf.

European Council, 1985. Council Resolution of 7 May 1985 on a new approach to technical harmonization and standards (85/C136/01). Official Journal of the European Union C 136, 4/6/1985: 1-9.

European Council, 1992. European Council in Edinburgh – 11 and 12 December 1992. Conclusions of the Presidency. DOC/92/8.

European Council, 2000a. European Council in Nice – 7, 8 and 9 December 2000. Conclusions of the Presidency. DOC/00/30.

European Council, 2000b. Council Resolution on the Precautionary principle. Annex III to the Presidency Conclusions of the Nice European Council Meeting, 7-9 December 2000. SN 400/00 ADD 1 (2000).

European Economic and Social Committee (EESC), 1998. Opinion on 'The single market and the protection of the environment: coherence or conflict (SMO)'. Official Journal of the European Union C 019, 21/1/1998.

European Food Safety Authority (EFSA), 2003. The effects of nitrites/nitrates on the microbiological safety of meat products. The EFSA Journal 14: 1-31.

European Food Safety Authority (EFSA), 2007. Scientific Opinion of the Panel on Animal Health and Welfare on a request from the Commission on the animal welfare aspects of killing and skinning of seals. The EFSA Journal 610: 1-122.

European Food Safety Authority (EFSA), 2010. EFSA scientific colloquium 13, Summary report what's new on novel foods. Amsterdam, the Netherlands: 19-20 November 2009.

European Parliament, 1997. Report on the alleged contraventions or maladministration in the implementation of Community law in relation to BSE, without prejudice to the jurisdiction of the Community and the national courts, A4-0020/97/A, PE 220.544/fin/A.

Food and Agriculture Organization of the United Nations (FAO), 1997. Risk Management and Food Safety. Report of a Joint FAO/WHO Consultation – Rome, 27-31 January 1997. FAO Food and Nutrition Paper 65. FAO, Rome, Italy.

Food and Agriculture Organization of the United Nations/World Health Organization (FAO/WHO), 2002. Report of the Evaluation of the Codex Alimentarius and Other FAO and WHO Food Standards Work. FAO, Rome, Italy. Available at: http://www.who.int/foodsafety/codex/eval_report/en/.

Food and Agriculture Organization of the United Nations/World Health Organization (FAO/WHO), 2006. Understanding the Codex Alimentarius. FAO, Rome, Italy.

Fisher, E., 2007. Risk regulation and administrative constitutionalism. Hart, Portland, OR, USA.

Fisher, E., 2009. Opening Pandora's box: contextualising the precautionary principle in the European Union. In: M. Everson and E. Vos (eds.), Uncertain risks regulated. Routledge-Cavendish, Abingdon, UK, pp. 21-45.

Food Standards Agency (FSA), 2007. Guidance notes for food business operators on food safety, traceability, product withdrawal and recall. July 2007. UK Food Standards Agency, London, UK. Available at: www.food.gov.uk/multimedia/pdfs/fsa1782002guidance.pdf.

Food Standards Agency (FSA), 2009. Advice from the advisory committee on novel foods and processes 2006. UK Food Standards Agency, London, UK. Available at: www.foodstandards.gov.uk/news/newsarchive/2006/sep/gmricetest.

Foster, K.R., Vecchia, P. and Repacholi, M.H., 2000. Science and the Precautionary principle. Science 228: 974-981.

Franch Saguer, M., 2002. La seguridad alimentaria: las agencias de seguridad alimentaria. Revista de Administración Pública 159: 315-340.

Freestone, D. and Hey, E., 1996. Implementing the precautionary principle: challenges and opportunities. In: D. Freestone and E. Hey (eds.), The precautionary principle and international law. The challenge of implementation. Kluwer Law International, The Hague, the Netherlands.

Funtowicz, S.O. and Ravetz, J.R., 1993. Science for the post-normal age. Futures 25: 739-755.

Giddens, A., 1999. Risk and responsibility. Modern Law Review 62: 1-10.

Giddens, A. and Pierson, C., 1998. Conversations with Anthony Giddens: making sense of modernity. Polity Press, Cambridge, UK.

Goldstein, B. and Carruth, R.S., 2004. The precautionary principle and/or risk assessment in World Trade Organization decisions: a possible role for risk perception. Risk Analysis 24: 491-499.

Graham, J.D. and Hsia, S., 2002. Europe's precautionary principle: promise and pitfalls. Journal of Risk Research 5: 371-390.

Grey, P., 1990. Food law and the internal market: taking stock. Food Policy 15: 111-121.

Grey, P., 1993. Subsidiarity and EC food law. Food Control 4: 61-66.

Grigorakis, K., 2006. Ethics in food safety. In: P. Luning, F. Devlieghere and R. Verhé (eds.), Safety in the agri-food chain. Wageningen Academic Publishers, Wageningen, the Netherlands, pp. 647-677.

Griller, S., 2000. Judicial enforceability of WTO law in the European Union. Annotation to Case C-149/96, *Portugal* v. *Council*. Journal of International Economic Law 3: 441-473.

Grossman, M.R., 2006. Animal identification and traceability under the US national animal identification system. Journal of Food Law and Policy 2: 231-315.

Grundmann, R. and Stehr, N., 2003. Social control and knowledge in democratic societies. Science and Public Policy 30: 183-188.

Hanf, D., 2001. Flexibility clauses in the Founding Treaties, from Rome to Nice. In: B. De Witte, D. Hanf and E. Vos (eds.), The many faces of differentiation in EU Law. Intersentia, Antwerp, Belgium, pp. 4-26.

Harlow, S.D., 2004. Science-based trade disputes: a new challenge in harmonizing the evidentiary systems of law and science. Risk Analysis 24: 443-447.

Hervey, T.K., 2001. Regulation of genetically modified products in a multi-level system of governance: science or citizens? Reciel 10: 321-333.

References

Hinrichs, C.C., 2000. Embeddedness and local food systems: notes on two types of direct agricultural market. Journal of Rural Studies 16: 295-303.

Hood, C., Rothstein, H. and Baldwin, R., 2001. The government of risk: understanding risk regulation regimes. Oxford University Press, Oxford, UK.

Howse, R., 2000. Democracy, science, and free trade: risk regulation on trial at the World Trade Organization. Michigan Law Review 98: 2329-2357.

Jans, J.H., 2000. Proportionality revisited. Legal Issues of Economic Integration 27: 239-265.

Jasanoff, S., 1994. The fifth branch: science advisers and policymakers. Harvard University Press, Cambridge, MA, USA.

Joerges, C., 1997. Scientific expertise in social regulation and the European Court of Justice. In: C. Joerges, K.H. Ladeur and E. Vos (eds.), Integrating scientific expertise into regulatory decision-making: national traditions and European innovations. Nomos, Baden-Baden, Germany.

Joerges, C., 2006. Free trade with hazardous products? The emergence of transnational governance with eroding state government. EUI Working Paper 2006/5.

Joerges, C. and Neyer, J., 2003. Politics, risk management, World Trade Organisation governance and the limits of legalisation. Science and Public Policy 30: 219-225.

Jones, J. and Bronitt, S., 2006. The burden and standard of proof in environmental regulation: the precautionary principle in an Australian administrative context. In: E. Fisher, J. Jones and R. Von Schomberg (eds.), Implementing the precautionary principle: perspectives and prospects. Edward Elgar, Cheltenham, UK, pp. 137-160.

Kapteyn, P.J.G. and VerLoren van Themaat, P., 1998. Introduction to the law of the European Communities: from Maastricht to Amsterdam. Kluwer Law International, London, UK.

Klint Jensen, K., 2006. Conflicts over risks in food production: a challenge for democracy. Journal of Agricultural and Environmental Ethics 19: 269-283.

Knodt, M., 2004. International embeddedness of European multi-level governance. Journal of European Public Policy 11: 701-719.

Knudsen, I., Søborg, I., Eriksen, F., Pilegaard, K. and Pedersen, J., 2008. Risk management and risk assessment of novel plant foods: concepts and principles. Food and Chemical Toxicology 46: 1681-1705.

Krämer, L., 1984. EEC action in regard to consumer safety, particularly in the food sector. Journal of Consumer Policy 7: 473-485.

Krapohl, S., 2005. Thalomide, BSE and the single market: a historical-institutionalist approach to regulatory regimes in the European Union. EUI Working Paper LAW No. 2005/03.

Kreher, A., 1996. The new European agencies. EUI Working Paper RSC 96/49. Florence, Italy.

Kreher, A. and Martines, F., 1996. Le 'agenzie' della Comunita Europea: un approcio nuovo per l'integrazione amministrativa? Rivista Italiana di Diritto Pubblico Comunitario 1996: 97-118.

Kuijper, P.J. and Bronckers, M., 2005. WTO law in the European Court of Justice. Common Market Law Review 42: 1313-1355.

Lafond, F.D., 2001. The creation of the European Food Authority. Institutional implications of risk regulation. Notre Europe, European Issues 10. Available at: http://www.notre-europe. eu/uploads/tx_publication/Probl10-en.pdf.

Levi-Faur, D., 2010. Regulation and regulatory governance. Jerusalem papers in regulation and governance. Working Paper No. 1 February 2010. The Hebrew University, Jerusalem, Israel.

Levidow, L., Carr, S. and Wield, D., 2005. European Union regulation of agri-biotechnology: precautionary links between science, expertise and policy. Science and Public Policy 32: 261-276.

Liberatore, A. and Funtowicz, S., 2003. 'Democratising' expertise, 'expertising' democracy: what does this mean, and why bother? Science and Public Policy 30: 146-150.

Löfstedt, R.E., 2005. Risk management in post-trust societies. Palgrave Macmillan, New York, NY, USA.

Löfstedt, R.E. and Vogel, D., 2001. The changing character of regulation: a comparison and Europe and the United States. Risk Analysis 21: 399-416.

Luhmann, N., 1993. Risk: a sociological theory. De Gruyer, Berlin, Germany.

Majone, G., 1996. Regulating Europe. Routledge, London, UK.

Majone, G., 2002. What price safety? The precautionary principle and its policy implications. Journal of Common Market Studies 40: 89-109.

Majone, G., 2003. Foundations of risk regulation: science, decision-making, policy learning and institutional reform. In: G. Majone (ed.), Risk regulation in the European Union: between enlargement and internationalization. EUI Robert Schuman Centre for Advanced Studies, Florence, Italy, pp. 9-32.

Mandeville, B., 2003. L'Autorité européenne de sécurité des aliments: un élément clef de la nouvelle législation alimentaire européenne. Rivista di Diritto Agrario 2: 142-151.

Masson-Matthee, M., 2007. The Codex Alimentarius Commission and its standards. An examination of the legal aspects of the Codex Alimentarius Commission. T.M.C. Asser Press, the Hague, the Netherlands.

McKone, T.E., 1996. Overview of the risk analysis approach and terminology: the merging of science, judgment and values. Food Control 7: 69-76.

McNelis, N., 2001. The role of the judge in the EU and WTO. Journal of International Economic Law: 189-208.

Mendez, M., 2004. The impact of WTO rulings in the Community legal order. European Law Review 29: 517-529.

Meuwese, A., 2008. Impact assessment in EU lawmaking. Kluwer Law International, Alphen aan den Rijn, the Netherlands.

Miller, D., 1999. Risk, science and policy: definitional struggles, information management, the media and BSE. Social Science and Medicine 49: 1239-1255.

Millstone, E., 2007. Can food safety policy-making be both scientifically and democrtatically legitimated? If so, how? Journal of Agricultural and Environmental Ethics 20: 483-508.

Millstone, E. and Van Zwanenberg, P., 2001. Politics of expert advice: lessons from the early history of the BSE saga. Science and Public Policy 28: 99-112.

Mol, A. and Bulkeley, H., 2002. Food risks and the environment: changing perspectives in a changing social order. Journal of Environmental Policy and Planning 4: 185-195.

Mortelmans, K., 2002. The relationship between the treaty rules and community measures for the establishment and functioning of the internal market – towards a concordance rule. Common Market Law Review 39: 1303-1346.

National Research Council (NRC), 1983. Risk assessment in federal government: managing the process. National Academy Press, Washington, DC, USA.

Organisation for Economic Cooperation and Development (OECD), 2003. Emerging risks in the 21st century: an agenda for action. OECD, Paris, France.

Onida, M., 2006. The practical application of Article 95(4) and 95(5) EC Treaty: what lessons can we learn about the division of competences between the EC and Member States in product-related matters? In: M. Pallemaerts (ed.) EU and WTO law: how tight is the legal straitjacket for environmental product regulation? VUB Brussels University Press, Brussels, Belgium.

Otsuki, T., Wilson, J.S. and Sewadeh, M., 2001. A race to the top? A case study of food safety standards and African exports. The World Bank Policy Research Working Paper WPS2563. World Bank, Washington, DC, USA.

Palau, A., 2009. The Europeanization of the Spanish food safety policy: framing and institutional change and its relation to European soft law. 5th ECPR General Conference, 10-12 September 2009, Potsdam, Germany.

Peczenik, A., 2009. On law and reason. Springer, Berlin, Germany.

Peel, J., 2004. Risk regulation under the WTO SPS agreement: science as an international normative yardstick? Jean Monnet Working Paper 02/04. NYU School of Law, New York, NY, USA.

Pescatore, P., 1987. Some critical remarks on the 'single European act'. Common Market Law Review 24: 9-18.

Pieterman, R., 2001. Culture in the risk society. an essay on the rise of a precautionary culture. Zeitschrift für Rechtssoziologie 22: 145-168.

Post, D.L., 2006. The precautionary principle and risk assessment in international food safety: how the World Trade Organization influences standards. Risk Analysis 26: 1259-1273.

Radaelli, C.M. and Meuwese, A., 2010. Hard questions, and equally hard solutions? Explaining proceduralisation in impact assessment. West European Politics 33: 136-153.

Ratzan, S.C. (ed.), 1998. Mad cow crisis: health and the public good. NYU Press, New York, NY, USA.

Renn, O., 2006. Risk governance: towards an integrative approach. International Risk Governance Council, Geneva, Switzerland.

Rippe, K.P., 2000. Novel foods and consumer rights: concerning food policy in a liberal state. Journal of Agricultural and Environmental Ethics 12: 71-80.

Rijksinstituut voor Volksgezondheid en Mileu/Institute of Food Safety (RIVM-RIKILT), 2009. Risicobeoordeling inzake aanwezigheid van LLRice 601 en geïmporteerde rijst 2006. Available at: www2.vwa.nl/portal/page?_pageid = 35,1554101&_dad = portal&_schema = PORTAL&p_file_id = 12446.

Roberts, D., 1998. Preliminary assessment of the effects of the WTO agreement on sanitary and phytosanitary trade regulations. Journal of International Economic Law 1: 377-405.

Rogers, M.D., 2001. Scientific and technological uncertainty, the precautionary principle, scenarios and risk management. Journal of Risk Research 4: 1-15.

Santer, J., 1997. Speech by Jacques Santer, President of the European Commission, to Parliament on 18 February 1997. Bulletin EU 1/2 1997.

Scientific Steering Committee (SSC), 2000. First report on the harmonisation of risk assessment procedures – Part 1: the report of the Scientific Steering Committee's Working Group on Harmonisation of Risk Assessment Procedures in the Scientific Committees advising the European Commission in the area of human and environmental health. 26-27 October 2000. Brussels, Belgium.

Scientific Steering Committee (SSC), 2003. Final report on setting the scientific frame for the inclusion of new quality of life concerns in the risk assessment process. 10-11 April 2003. Brussels, Belgium.

Seiler, H., 2002. Harmonised risk based regulation – a legal viewpoint. Safety Science 40: 31-49.

Selznick, P., 1985. Focussing organizational research on regulation. In: R.G. Noel (ed.), Regulatory policy and the social sciences. The University of California Press, Berkeley, CA, USA, pp. 363-367.

Skogstad, G., 2001. The WTO and food safety regulatory policy innovation in the European Union. Journal of Common Market Studies 39: 485-505.

Slotboom, M., 1999. The hormones case: an increased risk of illegality of sanitary and phytosanitary measures. Common Market Law Review 36: 471-491.

Slotboom, M., 2003. Do public health measures receive similar treatment in European Community and World Trade Organization Law? Journal of World Trade 37: 553-596.

Slovic, P., 1987. Perception of risk. Science 236: 280-285.

Slovic, P., 2000. Perception of risk. In: P. Slovic (ed.), The perception of risk. Earthscan, London, UK, pp. 220-231.

Slovic, P., Finucane, M.L., Peters, E. and MacGregor, D.G., 2004. Risk as analysis and risk as feelings: some thoughts about affect, reason, risk, and rationality. Risk Analysis 24: 311-322.

Snyder, F., 2003. The gatekeepers: the European courts and WTO law. Common Market Law Review 40: 313-367.

Snyder, F., 2006. Toward an International law for adequate food. In: A. Mahiou and F. Snyder (eds.), La sécurité alimentaire/Food security and food safety. Martinus Nijhoff, The Hague, the Netherlands, pp. 79-163.

Szajkowska, A., 2009. From mutual recognition to mutual scientific opinion? Constitutional framework for risk analysis in EU food safety law. Food Policy 34: 529-538.

Szajkowska, A., 2010. The impact of the definition of the precautionary principle in EU food law. Common Market Law Review 47: 173-196.

Tinbergen, J., 1965. International economic integration. Elsevier, Amsterdam, the Netherlands.

Trachtman, J.P., 1998. Trade and ... problems, cost-benefit analysis and subsidiarity. European Journal of International Law 9: 32-85.

Trute, H.-H., 2003. From past to future risk – from private to public law. European Review of Public Law 15: 73-103.

Türk, A.H., 2006. The concept of legislation in European Community law. Kluwer Law International, the Hague, the Netherlands.

Ugland, T. and Veggeland, F., 2006. Experiments in food safety policy integration in the European Union. Journal of Common Market Studies 44: 607-624.

United Nations Institute for Social Development (UNRISD), 2004. Technocratic policy making and democratic accountability. UNRISD Research and Policy Brief 3.

United States Department of Agriculture (USDA), 2009. Statement by Agriculture Secretary Mike Johanns regarding genetically engineered rice 2006. Available at: www.usda.gov/wps/portal/!ut/p/_s.7_0_A/7_0_1RD?printable=true&contentidonly=true&contentid=2006/08/0307.xml.

Van Asselt, M. and Renn, O., 2011. Risk Governance. Journal of Risk Research 14: 431-449.

References

Van Asselt, M. and Vos, E., 2005. The precautionary principle in times of intermingled uncertainty and risk: some regulatory complexities. Water Science and Technology 52: 35-41.

Van der Haegen, T., 2003. The European Union view of the precautionary principle in food safety. American Branch of the International Law Association, New York, NY, USA 23-25 October 2003.

Van der Meulen, B., 2006a. Haalt de Warenwet 2007? Journaal Warenwet 7(3): 6-14.

Van der Meulen, B., 2006b. Haalt de Warenwet 2007? Deel II implementatie van Verordening 178/2002. Journaal Warenwet 7(4): 6-15.

Van der Meulen, B., 2009. Reconciling food law to competitiveness: report on the regulatory environment of the European food and dairy sector. Wageningen Academic Publishers, Wageningen, the Netherlands.

Van der Meulen, B., 2010a. The global arena of food law: emerging contours of a meta-framework. Erasmus Law Review 3: 217-240.

Van der Meulen, B., 2010b. Prior authorisation schemes: trade barriers in need of scientific justification. European Journal of Risk Regulation 4: 465-471.

Van der Meulen, B. and Van der Velde, M., 2008. European food law handbook. Wageningen Academic Publishers, Wageningen, the Netherlands.

Vogel, D., 1995. Trading up: consumer and environmental regulation in a global economy. Harvard University Press, Cambridge, MA, USA.

Vogel, D., 2001. Ships Passing in the Night: The Changing Politics of Risk Regulation in Europe and the United States. RSCAS Working Paper No. 2001/31.

Von Moltke, K., 1996. The relationship between policy, science, technology, economics and law in the implementation of the precautionary principle. In: D. Freestone and E. Hey (eds.), The precautionary principle and international law. The challenge of implementation. Kluwer Law International, the Hague, the Netherlands.

Vos, E., 1999. Institutional frameworks of community health and safety regulation: committees, agencies and private bodies. Hart, Oxford, UK.

Vos, E., 2000a. EU food safety regulation in the aftermath of the BSE crisis. Journal of Consumer Policy 23: 227-255.

Vos, E., 2000b. European administrative reform and agencies. EUI Working Paper RSC 2000/51. Florence, Italy.

Vos, E., 2001. Differentiation, harmonisation and governance. In: B. De Witte, D. Hanf and E. Vos (eds.), The many faces of differentiation in EU Law. Intersentia, Antwerp, Belgium.

Vos, E., 2005. Regional integration through dispute settlement: the European Union experience. Maastricht Faculty of Law Working Paper 2005/7, Maastricht, the Netherlands.

Wiener, J.B. and Rogers, M.D., 2002. Comparing precaution in the United States and Europe. Journal of Risk Research 5: 317-349.

World trade Organisation (WTO), 2000. Summary report on the SPS risk analysis workshop 19-20 June 2000. G/SPS/GEN/209 (3 November 2000).

Zonnekeyn, G., 2001. The latest on indirect effect of WTO Law in the EC Legal order. The Nakajima case law misjudged? Journal of International Economic Law 4: 597-608.

Zonnekeyn, G., 2004. EC liability for non-implementation of WTO dispute settlement decisions – are the dice cast? Journal of International Economic Law 7: 483-490.

About the Author

Anna Szajkowska was born on 7 January 1979 in Warsaw, Poland. She studied Law and Administration and French at Warsaw University and Agricultural Economics at the Warsaw University of Life Sciences. She also completed a programme on French and European Union Law organised by the University of Poitiers in co-operation with Warsaw University. Before joining the PhD programme at Wageningen University in 2005, she worked at the Ministry of Agriculture and Rural Development in Poland. In 2005 she received a scholarship for young foreign lawyers from the Italian National Research Council at the Institute of International and Comparative Agrarian Law in Florence to establish her thesis research. From 2006 to 2009 she worked as an international project administrator at the Centre for Water and Climate, Wageningen University. She is currently a researcher at the Law and Governance Group at Wageningen University. Her research interests include food law, EU law, risk regulation, the relation between EU law and national legal systems, and international trade law.

Printed in the United States
by Baker & Taylor Publisher Services